Building Design for Energy Economy

Building Design for Energy Economy

The Ove Arup Partnership

THE CONSTRUCTION PRESS
LANCASTER LONDON NEW YORK

This volume is based on a teaching seminar given by staff of the
Ove Arup Partnership in 1979.

The Construction Press Ltd, Lunesdale House, Hornby, Lancaster, England.

A subsidiary company of Longman Group Ltd, London.
Associated companies, branches and representatives throughout the world.
Published in the United States of America by Longman Inc, New York.

First published 1980.

British Library Cataloguing in Publication Data

Ove Arup and Partners
   Building design for energy economy.
   1. Architecture and energy conservation
   I. Title
   721   NA2542.3

   ISBN 0-86095-850-7

Printed in Great Britain by The Pitman Press, Bath.

Contents

CONTRIBUTORS

Thomas Barker, *Director, Ove Arup & Partners*

John Campbell, *Technical Director, Mechanical & Electrical Development Group, Ove Arup Partnership*

David Clarson, *Quantity surveyor, Arup Associates*

Peter Dunican, *Chairman, Ove Arup Partnership*

Robert Emmerson, *Director, Ove Arup & Partners*

Des Gurney, *Director, Ove Arup & Partners*

James Hampson, *Director, Ove Arup & Partners*

Michael Holmes, *Mechanical & Electrical Development Group, Ove Arup Partnership*

James Knight, *Energy consultant*

Tony Marriott, *Director, Arup Associates*

Clary McDonald, *Architect, Arup Associates*

Richard Robinson, *Energy Manager, John Player Ltd*

John Sinnett, *Director, Ove Arup & Partners*

Henk Spoormaker, *Partner, Ove Arup & Partners South Africa*

Soji Thomas, *Partner, Ove Arup & Partners Nigeria*

Peter Warburton, *Director, Arup Associates*

Chapter 1

Introduction

P Dunican

The papers reproduced in the following chapters were written for discussion at an internal seminar of The Arup Partnerships. As a consequence there are some domestic references which may not be immediately clear but which the context should clarify.

The notion of the seminar stemmed from a belief that the structural engineer for the time being, by virtue of his background, experience, and technological role in the building design team, could make an effective contribution in designing for energy economy during the formative stage of a building scheme. It is a question of developing his understanding of the physics of building and of helping and guiding the architect as the design leader to assess the energy implications of changes in the siting, the form and the construction of the building; to help bridge the gap of understanding between the architect and the services engineer.

It has become commonplace to argue that buildings should be designed for energy economy, whether it be for heating, cooling, lighting, or using, but in practice still too little is done about it. The architect may understand the generalities of the matter but is he as well equipped as he should be to assess either qualitatively or quantitatively the effect that changes in his design will have on the energy needs of the building? No, he is not and neither is the services consultant, if he is more concerned with the detailed design of engineering systems to cope with the problems posed by preconceived solutions rather than consideration of the total energy problem and its equitable solution.

Given some education, guidance, and training the structural engineer could in the meantime make a valuable contribution to this very real problem as an intermediary or entrepreneur, if he has the will to do so.

In the long term, however, we are seeking a source of inspiration. In Britain in the thirties the practice of structural engineering was intellectually at a relatively low level. It was concerned mainly with the analysis, calculation, and construction of structural frameworks usually conceived by others, generally architects through their planning of buildings, using some rules of thumb and their individual experiences. Later, as younger designers became interested in the use of reinforced concrete, structural conception as it exists today began to emerge. This important modern movement was led by a few pioneers such as Ove Arup and Felix Samuely. They helped to transform structural engineering from a trade, albeit skilled, into a creative profession and thereby established the significance of the structural engineer's role in the building design team and the design of buildings. This transformation was aided by some architects whose understanding of structural behaviour and possibilities was exceptional.

As a sweeping generalisation I think that today building services engineers are in the position in which structural engineers were before the war. They have the science and technology but they need more help in its application — in the art of building services engineering. It is only a matter of time before this situation is dealt with through the emergence of some inspirational personalities and individual challenges, and also, but perhaps less likely, a significant growth in the architect's understanding about the building services problems and their possible solutions within the context of energy use and conservation.

Meanwhile, we have tumbling down upon us the imperative demand for energy economy in buildings, which needs much more than bigger and better systems, if it is to be met. How, therefore, can we best respond now to our present needs when for possibly the next decade or so the most natural advisers, the building services engineers, will still be acquiring the necessary deeper understanding of the potentials of their science and technology, particularly as it applies to the way people and buildings behave together?

We can return to one positive suggestion that in the meantime to help bridge the gap it is for the structural engineer, who at the moment knows something about the physics of the building, to be encouraged to contribute more to the solution of the energy problem than he presently does at the conceptual stage of the architect's thinking about the building problem.

To discuss and develop this idea is the primary purpose for which the earlier mentioned seminar was set up and the papers written. They are not intended to be the last word on the subject, but they are intended to contribute to better buildings.

Chapter 2

Energy and buildings – how it is used

J Knight

All energy used in creating the microclimate in buildings is wasted. It disappears through the windows and fabric of the building and is exhausted to the outdoors as vitiated air. When, however, energy is used in manufacture or agriculture there is an end product, however wasteful the process of making it may be. It is this vital difference in the result of using energy which ought to be driving us all towards the maximum economy in using it. Nearly everybody is motivated, sadly, by self-interest and the cost and tariff structure of fuel and power supplies has not encouraged the study of ways and means to reduce consumption and the suppliers of energy have been motivated only by the desire to sell as much as possible.

Increasing the price of energy to encourage conservation is not an attractive policy to many people, as at first sight it appears inflationary and penalises the poor and elderly. But a letter to the Chartered Mechanical Engineer, Journal of the Institute of Mechanical Engineers, published in the February 1979 issue, offers a solution which far from being inflationary would seem to have the reverse effect. Professor D G Wilson of the Massachusetts Institute of Technology writes in favour of a new policy on pricing fuels and electricity which, to put it very roughly, argues for the current prices to be weighted with the external costs of winning the energy sources, arising from pollution, injury to workers, transportation, treatment and storage of nuclear wastes, and the research and development work necessary. Also to be added to the price is the effect on our balance of payments of buying foreign fuels and the costs of Defence forces in the safeguarding of suppliers and routes of supply. All these costs, it is argued, are borne by the average taxpayer and so concealed from the fuel user. Another factor to be added is the cost of replacing depleted resources. The amount paid for energy would be lower for the fuels not incurring these costs (ie, the safer ones) and taxes could be reduced accordingly.

In the very unlikely situation of a government and the energy suppliers instituting such a costing system, the real cost of energy would become apparent to all and in our own self-interest great strides would be taken towards real energy economy and self-sufficiency. In the meantime, as everyone knows, about 50% of our total energy consumption is used in buildings. Only 40% of our entire electricity production goes to industry and agriculture. We use more electricity in heating our homes than any other Western country, not because we have more electrically heated homes but because of our wasteful use of it, and lack of thermal insulation.

Not enough time has elapsed since the so-called energy crisis of 1973/74 to see whether the rate of growth of consumption has been altered, but prior to the crisis it was estimated that the rate of 5% per annum would be maintained into the 1980s. In the USA it was thought that the growth rate would mean doubling consumption in 14 years, even though consumption there is double that of Europe in proportion to population. Although energy consumption is usually proportional to gross national product, we in the UK use considerably more than the more productive countries of the EEC. The figure for Europe is around 5 kW/person/day and in the USA about 10 kW/person/day. It is said that we need food intake equivalent to 0.15 kW/day. Each person in the industrialised world has,

9

at his elbow, the equivalent of 30 human slaves. If the entire world consumed energy at the American rate we would need to obtain over six times the present amount used — a quite impossible situation.

Fifty per cent of our total energy supplies used in building can be broken down as follows:

Building industry — winning and mining materials	5%
Heating and ventilation	30
Domestic hot water	3
Cooling and humidity control	2
Electric lighting, TV, radio, water and sewage	8
Transportation between living area and work	2
Total	50%

The proportions will vary from country to country but seem to be representative of Europe. Transportation as a whole takes around 17% of total energy demand and in this context it is interesting to note that a single car occupant travelling 15 miles and back to work uses 2–3 times the amount of energy he uses in his office.

When we consider how energy is used we must bear in mind that there are something like 50 000 major buildings in the public sector and rather more than 100 000 in the private sector. Central and local government offices, Departments of Health, Education and Defence together consume 330×10 GJ/annum while shops, warehouses, hotels, commercial offices, bus and train termini about 20% less, in terms of fossil fuels. Shops and warehouses use a higher proportion of electrical energy than any other single group, no doubt due to high levels of lighting and advertising signs, etc. Of all the classes of users in the public sector hospitals and the like have the highest demand per square metre of floor area.

The normal rate of growth of new building in Europe appears to lie between 3 and 5 per cent per annum, so clearly no real impact on reduction of fuel and power consumed can be made in the short term without tackling the existing stock of buildings. A recent survey of the state of existing buildings in the USA revealed that 50% are over ten years old and the average building over that age wastes 40% of the energy consumed due to out-of-date boilers and firing systems, faulty fans and refrigeration plants and unlagged pipes. In those buildings lighting accounted for 30–50% of the cooling load. Of course the figures are not representative of UK conditions, but the proportion of old buildings here is no doubt greater; perhaps 90% would be a fair figure. The application of optimum start controls to the Property Services Agency buildings shows that 30–40% of fuel used in the heating systems could quite easily be saved and similar results are available from some of the large commercial organisations. It is interesting to note that a recent analysis of US consultants' work load showed that on an average 40% was on energy audits and conservation studies generally.

In a later chapter there is discussion of the building envelope and its effect on energy use but as yet it is uncertain as to whether there is enough data on actual buildings of like characteristics to allow comparisons to be made.

Comparisons have been made of the theoretical energy consumption of slab-type offices as against the large open plan designs and the impression given that because the latter require artificial cooling of the inner core then they are inevitably more greedy for energy. Refrigeration, even in much warmer climates than our own, usually accounts for only about 15% of energy used and I feel sure that this can greatly be reduced in the UK by better design, selection of higher chilled water temperatures and lower condensing temperatures than hitherto customary. Also, as a rule we do not go to sufficient trouble to recover waste heat from the condensing system. In large installations raising chilled water temperature by a couple of degrees and lowering the condensing temperature by a similar amount can save 40% of power consumed. The British climate provides many opportunities to air conditioning designers to save by breaking with traditional practices and standards in this way.

We should certainly now be pressing all manufacturers to provide certificates of energy used by their equipment, as it is not unusual to find very wide differences between one make and another. The work of checking manufacturers' submissions for the main items can become tedious and time-consuming, but should be a vital part of the consultant's service to the client.

ALTERNATIVE ENERGY SOURCES

It is unwise to make forecasts of the great contribution to energy resources which will be made if only we spend more on research and development in the fields of solar radiation, wind power and the use of the tides. Whatever contribution will be made in the future it will not occur for a very long time. We have the record of nuclear power development to guide us in appreciating the time-lag between promise and achievement. At the present time there must be well over a hundred firms making or installing flat plate solar collectors for domestic hot water supply. It is very difficult to see how such activity is justified in this country, especially when it is appreciated that 50% of the energy used to provide hot water is wasted on standing losses. Unfortunately, the materials currently being used are highly energy intensive and all subject to damage by the weather over quite a short period, so it is likely that by the time an installation has worked long enough to start to make some return on capital it will need replacing. Be that as it may, it is reliably reported that if all the possible installations were built in the UK the maximum contribution in a good year to our energy resources would be less than 1%. No account in this estimate is made of energy used in their manufacture. Other countries and other climates are a different matter. At present the EEC spends £5.6 millions in total on solar energy research each year, but the scientist in charge thinks it unlikely that even 3% of Europe's energy needs will be met in this way in the year 2000.

Wave energy is another possible source of electrical energy but the most recent estimates indicate that the cost of the structures necessary will be between £4000 and £9000 per kW installed, meaning that the cost to the user cannot be less than 20–50p kWh. Conventional fossil fuelled or nuclear power stations cost £500–£1000 installed kW of capacity at a unit cost of electricity to the consumer of about 2p.

Too little is known about the costs of producing electricity on a large scale by wind power to allow even tentative forecasts to be made, but the capital costs are likely to approach those of the wave power machines.

We have to compare these costs and the ecological problems associated with the benefits to be derived from combined heat and power production from existing power stations where the efficiency of the plants could be raised from a maximum average of 30% to 70–80%. Nearly all our big cities could be served in this way, as are the city centres of New York, Pittsburg, Baltimore etc. Steam from the generating turbines is used to drive the enormous refrigeration plants in the new Citicorp Building, as is the case in other similar buildings. In most of the eastern European countries and in Denmark and Sweden combined heat and power production is the norm. The savings in energy to be achieved in this way far outweigh any that may eventually be obtained from the previously mentioned alternative sources, so one wonders why we do not get down to the job of doing it instead of continually talking about it. Some clue as to the reason why may be gleaned from the reports of the recent conference organised by the Electricity Council entitled "Energy Effectiveness in Buildings". The Council's Energy Sales Manager argued that building designers should now be designing for dual fuel systems so that they may quickly be converted to 'All Electric'. He also chastised the Department of Science and Education for issuing Design Notes for the Environmental Design of Educational Buildings which advised that all energy consumption comparisons should be in terms of 'Primary Energy Units'. It was said that the Council was appalled that such units are recommended and that they are a crazy yardstick. Of course, such a yardstick does show up the real efficiency of use in total energy terms and thus highlights the wasteful use of fuel in electricity-only power stations. But it will be noted that no mention is made of the need to plan for eventual combined heat and power production, although the Council is committed to carry out a development scheme in a major city.

Only the strongest pressure of public opinion is likely to bring about a new attitude on the part of the Central Electricity Generating Board and it is hoped that this issue will be an opportunity for all the professions involved in building to speak up. There is an alternative energy source available here which does not mean spending vast funds on research and development and the kind of time lag we have already seen occur with other schemes.

USE OF ENERGY INTENSIVE MATERIALS

Concentration upon fuel and power consumption by the engineering services had led to some neglect of the problems which will be presented in the future by our use of energy-intensive materials not available for recycling for several decades. Our designs have not hitherto been concerned to avoid their use for obvious reasons, but with the increasing demands by the third world which appears to be industrialising much more quickly than thought possible 10 or 15 years ago, it is likely that many of the materials we take for granted will become either extremely expensive or in short supply. More thought should be given to avoiding the unnecessary use of these when designing schemes. At the moment there does not seem much choice with the electrical installations, but there is a choice in the case of mechanical systems. For example, a modern VAV type air-conditioning system uses much less copper, stainless steels and the like than the older type of piped water and air induction systems and is more economical in capital and running costs. We should prepare for a future when materials need as much conserving as fuel and power.

Chapter 3

Building design

C McDonald

Much of what the building designer does depends on his philosophy — his understanding of why he is doing what he is doing. If he can identify the basic motivation behind his actions, he is in a better position to carry out the practical side of the work in a balanced manner. It is the achievement of this balance that gives the end result of his effort the properties associated with good design or architecture.

The fuel crisis of 1973 and the resulting escallation of fuel costs acted as the trigger which prompted Government and public awareness of the whole concept of energy usage. Designers are now actively being encouraged to look at energy usage and conservation in many different fields. At present their approach to this problem is very diverse and seems to lack any form of comprehensiveness. They find it difficult to decide if it means reduction in comfort standards, increases in capital costs, waiting for new legislation, or taking a more energy conscious attitude to designs.

What should designers' attitudes be? Obviously they should use their talents to a maximum in order to achieve an efficient use of energy, in solving the problems which society and clients put upon them. There lies the catch — what does an efficient use of energy mean and what approach can be adopted to achieve it?

Many of the preceding generation of buildings have the common characteristic of being comparatively high energy consumers. Energy was used, as it was then inexpensive, to overcome problems of heat loss, solar-heat gain, lighting, cooling, etc. Few architects, and in fact society in general, were not particularly aware of the serious implications of resource consumption. While it is reasonable to question why society as a whole was short-sighted, it must not be forgotten that buildings of this era represented the general social concerns of their time. In doing so they have solved many of the social and technical problems of our age, and still continue to do so. Designers developed the ability to house groups of people and activities in a way which we now take for granted. Therefore, although it is easy to condemn many of these buildings, in the light of our 'new found knowledge' we must do so in a manner which does not throw out the good with the bad.

It is equally interesting to look at buildings designed before energy became plentiful and easily distributed, as well as buildings from cultures where energy, as we use it, was not available. Designers of these buildings overcame many of the problems we face by the manipulation of such things as thermal mass, shading, orientation, wind, sun, etc. In addition, these devices have often been instrumental in forming the architectural vocabulary of the buildings to which they were applied. It is not suggested that these concepts will solve all the problems; however, they must not be disregarded.

"Why redesign the wheel when the man around the corner is already successfully using it"?

Many architects and designers today would lead one to believe that their 'art' has to be sacrificed in order to attain energy efficiency in buildings or alternatively that they have to sacrifice energy efficiency in order to attain 'art'. Neither of these statements should be regarded as being the truth. To examine why, one only has to look briefly at what architecture is.

Architecture is a complex undertaking involving technical, social, utilitarian and cultural problems. In short, it is the systematic arrangement of knowledge, it involves the making of beautiful forms, ordering space in a coherent manner, use of materials in a functional and knowledgeable way and providing comfort and convenience for its clients and their activities. To achieve architecture, the designer must be able to manipulate these diverse elements in a manner which will produce a balanced and ordered whole. If the architect singles out any one aspect of his craft to the detriment of the others he will, by definition, produce a flawed piece of architecture and will therefore have failed. Thus, in today's climate a concern for energy conservation must form a part of the designer's brief and must be included in the equation to the detriment of neither itself nor the other aspects of the brief.

The thesis is that the designer's immediate objective is to utilise all his knowledge of the past and present, combine it with the new technology in a balanced and compatible manner and thus produce a result that can be defined as architecture. He must not again make the mistake of pursuing one particular objective or technology to the detriment of the others. This seems an obvious statement; however, if we examine what is beginning to happen today we can see evidence of designers over-concentrating on energy economy to the detriment of other aspects of their craft. It is not suggested that designers disregard energy conservation. It is, however, put forward that they must apply it as an integral part of the total objectives of their professions.

To do this, these objectives have to become common objectives at the conceptual stage of any design process. Since the design of a building requires the active participation of a multitude of professions including a client, all of these contributors require to have as part of their vocabulary the concepts of energy economy. All designers have a basic feel for their fellow professionals' disciplines and intuitively take account of them in their design processes. They need to develop in a similar manner a manipulative ability of the concepts of energy economy in order that they can include the concepts in the basic equation of the design, rather than regard them as a subject which can be applied to the end process of a design.

The designer's philosophy must be to design for energy conservation in a comprehensive manner, taking account of all the multivarious aspects of his art. He must explore new technology and take account of past technology but always apply the results in a manner which produces a coherent whole.

There are many practical aspects of this process which can be examined. The most important and effective ones are obviously those which require to be considered at the conceptual stage of the design process.

Chapter 4

Financial aspects

D Clarson

When the financial aspects of energy conservation are considered, it is seen that many of the influences on what and how money is spent in the design of a building come from outside the design team.

In most cases the largest single influence is the client. His attitude can vary: he may have little awareness; he may have no real concern; he may not be prepared to spend any additional money; he may be committed to the idea; he may be prepared to spend additional money that will show a reasonable return. Therefore, the design team's objectives have to be set within the framework of the client's attitude and although we may wish to influence that attitude we must not over-estimate our ability to do so.

Public client bodies are generally encouraged to take a progressive view in energy conservation, but the attitude of many private clients may be less encouraging. It is important to remember that on average the energy costs for running our buildings represent 5% of our personal energy consumption and 2–3% of industrial process turnover (1).

For private clients energy costs are not a major factor in investment decisions. It is only when the 5 per cents and the 2–3 per cents are added together that they become significant in economic terms.

If the capital cost is the most important factor influencing a client and energy saving is seen as a hindrance in the early stages, consideration may be best given to energy saving devices that can be an 'optional extra'. At a later stage these can be considered in self-financing terms.

The trend of the last few years that is almost certain to continue is more building legislation that encourages and insists on energy conservation. So far this has not been too dramatic in design or cost terms. The effect of legislation on cost has been absorbed within the upward drift in building prices due to the demand for generally higher standards in building design.

Whether this gradual movement is intentional or not, it has been prudent as it has not caused any short-term distortion in the building market which would have a particularly pronounced effect during the present depressed period. The temptation may be for the legislators on energy conservation to wait for the next building boom when dramatic changes could become confused with the general upheaval of the building market.

Another trend which is almost certain to increase is the use of government grants and incentives to encourage energy saving. What should the attitude of designers of new buildings be to these? They may all be encouraged by (and have applied for) grants to insulate lofts, but does the specific grant system show an understanding of the design process and encourage their creativity? The fallacy of the grant or subsidy system is becoming well known and acknowledged in our society and coupled with its possible abuse it may not be as beneficial as may first appear.

An example of this misunderstanding is the suggestion by an MP that the housing cost yardstick be increased when a particularly low energy heating system is incorporated into the design of a house (2). Although the sentiment is exemplary, unfortunately it can encourage higher building costs to achieve energy savings.

The calculation of the cost effectiveness of energy conservation measures has so many variables that generalisations can be dangerous. It is important to understand the basis of calculation of any conclusion taken from another source, before presenting it to a particular client. It would be nice, even if unrealistic, if all clients would accept the simple calculation that if I spend £1000 more initially and save £200 per year I have a pay back period of 5 years.

For any detailed mathematical cost-effective calculations to be attempted a number of factors need to be known, or rather assumed, over and above the capital cost, energy cost and maintenance costs:

a) What return does the client expect from his money? (For many manufacturing processes this is a very short period.)

b) What general inflation rate is to be assumed?

c) Will energy costs (or prices fixed by Governments) outstrip the general level of inflation?

d) What tax allowances can the client claim on his various categories of expenditure?

Over the last 20 years Quantity Surveyors have become used to being involved in setting the level of the client's budget and indicating to the various design professions the amount of money available for particular parts of the project. These budgets are largely arrived at from historical data and money is often allocated throughout the building elements on a percentage basis. Arising from this the phrase 'a well balanced cost plan' has been added to our vocabulary and this is often used as a judge of the validity of a cost plan. The damage of this approach, particularly where designers work in water-tight compartments, is that there is little opportunity for flexibility in design approach to achieve energy efficiency.

Due to the Quantity Surveyors' reliance on historical data there is a problem in making a real and responsive contribution at an early stage when the designers are attempting an energy-conscious design strategy which does not follow the traditional pattern of expenditure on a building.

Other aspects which need early consideration are where the energy conservation demands a higher specification of the building element, eg, concrete structures which do not leak air, opening windows which do not leak, etc. Also, any change in traditional constructional sequences must be considered early on.

REFERENCES

1 Department of Energy. Energy Paper No 32, "Energy Conservation Research, Development and Demonstration: an Initial Strategy for Industry", The Department, 1978.
2 Allaun, F, "Homes Breakthrough", Building, *236* (5), p 41, 2 February 1979.

Chapter 5

Services energy effects

P Warburton

A contribution to reducing primary fuel consumption in buildings may well be possible where developing countries are expanding their policies for the generation of electricity and the distribution of fuels. If designers can affect government policy and know what to do, then they must try. Unfortunately, this is outside the designers' normal sphere of involvement in Britain and they can only aspire to some suggestions that it would be better to use the heat from electricity generating stations rather than throw it away. Perhaps the environmentalist would disagree and argue that electricity stations should be as far as possible from our energy-consuming cities and towns, which makes the distribution of the waste heat uneconomic.

A design firm's involvement in energy consumption occurs after a gas meter, electricity meter or lorry has delivered its fuel to it. Somebody else has dug and pumped the fuel, or generated electricity for it and the efficiencies are beyond its control. Fortunately, in Britain the efficiency of producing electricity is reflected in the cost difference between electricity and other fuels, so that a decision made on the basis of cost is also a decision that reflects the oil or coal consumption of the electricity generating system.

Before considering services machinery designers should try to establish how energy economy can contribute both to the conceptual form of a building and to the conceptual design of any services within that building.

The design of buildings for countries where energy is not readily available shows an awareness of climate, shading, wind and materials which has evolved through the process of trial and error to provide comfortable building enclosures. Where inexpensive energy is readily available, building services engineering is involved in using this energy to solve the problems created by building enclosures designed without this awareness. Recently, designers have developed much more accurate mathematical ways of analysing these problems and one contribution must be in the application of these techniques to designing comfortable building enclosures while at the same time minimising the use of non-renewable fuel through an awareness of climate, shading, wind and materials. This will reduce the need for an extended process of evolutionary trial and error, so that we can still maintain our standards of comfort before high fuel costs require us to reduce them. If designers accept this, then the windows, walls and construction of a building can be considered partly as services components and can be analysed before services machinery is considered.

In some developing parts of the world a building designer's client and the occupants of the buildings he designs may well expect standards of comfort based on European or American attitudes. These can rarely be achieved by local building designs and comfort standards evolved when energy is not readily available.

The evolutionary design of buildings is different for each culture and for each climate and it would not be right to propose one set of attitudes as a solution for all, but only to quote it as an example for discussion. In Britain there is an abundance of wind, rain and cloudy daylight. If the wind can be used to provide comfortable natural ventilation through a technical analysis of window design and at the same time the problems of summer overheating can be balanced with maximum daylighting and winter solar heating, then energy consumption is substantially reduced. The hours of use of artificial lighting will be minimised and the buildings will not require ventilating fans. But they will require heating and for energy economy we need to regularly recalculate and re-establish a higher standard of building insulation and to reduce and control more accurately the ventilation rates for buildings just to prevent condensation.

Assuming that for cultural, climatic, or local site reasons a building form is established that will require services machinery to make it comfortable, where can most energy economies be made? Designers are more likely to be successful if they start with the largest primary energy consumers, and that order will be different for different types of building. It will include artificial lighting, fan and pump drives, cooking equipment, cooling plant, heating plant and lifts. All have received much attention from designers but generally those items of equipment that run continuously at maximum during occupied hours consume the most.

Now assume that a building has been designed with the minimum amount of primary energy-consuming services machinery. The possibility of purchasing additional services machinery at extra cost to save energy can then be considered. These will include heat recovery, the use of cold outside air for cooling, heat pumps, on-site generation of electricity with heat recovery for heating and cooling (total energy). It is at this stage that some possibilities for the integration of built form and servicing equipment arise. For example, the relationship between thermal mass, the occupants and the daily temperature swings may be combined to reduce the fuel consumption required for summer time comfort cooling. If designers encourage the natural buoyance of warm air from people and machines and remove it effectively, they can reduce the fuel consumption required by reducing the temperature differences required to produce comfort cooling.

Let us reconsider the client's understanding of the implications of energy-saving measures. There is no point in installing complex services energy-saving machinery in a building if the client is not committed to understanding and then maintaining it. Maintenance costs must be debited against fuel savings as part of the cost analysis of energy saving options.

The subject of maintenance raises many questions for the designer and the building operator. When buildings are designed to an energy target, both the commissioning engineer and the maintenance engineer can prevent this being achieved. Recent surveys of groups of similar buildings indicate that the abnormal energy consumer occurs where commissioning or maintenance or both have been inadequately carried out. It is important, therefore, that energy targeting and planned maintenance are part of the building operator's accepted work.

Lastly, clients expect designers to be aware of new techniques which hopefully will one day prove their viability in the commercial world. Of these only solar energy has appeared on the British market in a form that can begin to satisfy the designers' cost analysis and then only in specific cases such as swimming pools. Unfortunately, the sun does not shine very often in Britain. If only windmills could be made to work efficiently!

Chapter 6

Microclimate

R Emmerson

INTRODUCTION AND DEFINITION

This chapter is intended only as a brief introduction to the subject and to act as a basis for more detailed discussions. Olgyay's "Design with Climate" represents an invaluable summary of the subject:

"We are inclined to think of climate as a certain condition uniformly distributed over a large area. This impression is partly because weather data are collected at points where 'undisturbed conditions' prevail and partly because large-scale maps depict equal mean temperatures in a few smooth lines. However, at ground level multifold minute climates exist side by side, varying sharply with the elevation of a few feet and within a distance of a mile. Nature demonstrates this ... the difference between the kind of plants which would grow on either side of a hill, if nature were allowed to select, would be as great as the difference between locales a hundred miles north and south of one another. Further, every elevation difference, character of land cover, every water surface induces variations in a local climate.

These effects within the large scale 'macroclimate' form a small scale pattern of micro-climates."

Therefore, microclimate will be defined as the climatic conditions local to the area under consideration, whether it is a large site or a small building. For example, one could talk of the microclimate of a 300 ha site or the microclimate around or within a building on that site and, obviously, these microclimates may and probably will be different.

DATA

The purpose

The range of data required is obviously a function of the proposed use of the site. However, the quality and quantity of data required have tended to increase as designs have increased in sophistication, and in the last two or three years have increased as analyses have become more sophisticated in an attempt to design 'climate moderated' or 'natural conditioned' buildings as opposed to 'mechanically conditioned' buildings.

Figure 1 Potential of climatic controls.

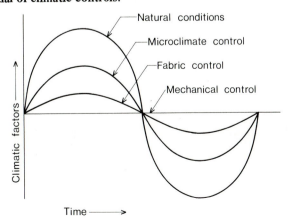

If we desire to work with, and not against, the forces of nature then we must have a better understanding of these forces, both on the macro-scale and the micro-scale. The study of site microclimate and its possible modifications is only the first step in a large and complex design process. Some other stages of this design process will be illustrated in later chapters.

Microclimate modification or control can significantly reduce the diurnal variation in climatic conditions, leaving proportionally less to be achieved by passive means, (ie, by appropriate selection of the building's form, orientation and the thermal properties, proportions and distribution of the facade). The fine tuning of the internal environmental conditions can then be achieved by active controls, ie, the engineering services installation (Fig 1).

Data collection*
(Tables 1–4)

Temperature
Dry bulb air. The usual form of presentation of this information is in monthly means (maxima and minima) and in monthly absolute figures (over, say, a 20-year period) of maxima and minima'.

Although this information may be adequate for determining design temperatures as per IHVE or ASHRAE, it is generally not good enough for our purposes. We need to know average hourly temperatures and their relation to other factors such as wind speed and direction, relative humidity, etc.

Soil temperatures. This information will only be required infrequently. However, in hot arid climates, where the moderating effects of nocturnal radiation are being considered, a detailed knowledge of the variation of soil temperature with depth over a 24-hour cycle is required. Obviously, data as to the nature of the soil and its albedo — the ratio of reflected radiation to total radiation — are required.

Water temperatures. Again, this information is required only when large bodies of water are adjacent to the proposed site.

Humidity and evaporation
Ideally, this should be in the form of average hourly percentages for humidity and average hourly rates from saturated materials for evaporation.

Rainfall
Required are average monthly intensities, monthly minimum and maximum, and 24-hour intensities, and 1-hour storm intensities.

This should be combined with some idea as to the run-off characteristics of the existing strata and potential catchment areas.

Wind
Available wind data are usually in the form of 10-minute mean recordings or 15-sec mean recordings observed either every hour or every three hours. As for temperature data, in many parts of the world the readily accessible data are in the form of monthly averages giving percentage of wind from various directions with average velocities, together with a table of absolute monthly maxima with directions. Ideally, one wants finer detail than this, preferably in the form of hourly 10-minute means. Three-second gust values, if not recorded, can be extracted from this information.

However, a word of caution: for tropical countries one should look at the source (population) of the data. For example, the mean monthly gust velocity data may be drawn from a mixed population of normal gusts and rarely occurring tropical gusts. The tropical storm gust will be vastly greater than the normal (non-tropical storm gust).

Insolation
Very little quantitative insolation data is available other than in the form of number of overcast, partially cloudy and clear days per month. An alternative form is the number of hours of sunshine per month. However, there is uncertainty in the method of translating the number of hours of sunshine to total radiation energy received. Finally, in some areas radiation data are available only on an annual basis, precluding the use of information for rational design of solar processes in areas where seasonal variability of radiation is large.

However, some radiation recording stations have been established for some time (notably Kew in the UK) and others have more recently been established.

---

*"Survey of meteorological information for architecture and buildings". BRS paper CP5/72.

Table 1 Meteorological information requirements for town planning.

	1/1	1/2	1/3	1/4	1/5	1/6	1/7	1/8	1/9	1/10	1/11	1/12
	Fog, interference with traffic, also pollution control	Radiation climate	Wind (cold stress)	Exposure to rain	Exposure to snow	Soil moisture	Drainage, risk of floods	Availability of illumination in buildings	Snow accumulation	Height control: winds at ground level	Height control: icing risks	Height control: low clouds
1 Solar radiance, direct on horizontal		*				*						
2 Solar radiance, diffuse on horizontal		*				*						
3 Illuminance direct on horizontal								*				
4 Illuminance diffuse on horizontal								*				
5 Sunshine duration		*						*				
6 Sunpaths		*										
7 UV radiance total on horizontal												
8 Equivalent radiative temperature of sky												
9 Temperature of air at standard height			*			*				*	*	
10 Temperature of air at vertical variation	*										*	
11 Temperature in ground												
12 Humidity of air	*					*						
13 Fog	*										*	
14 Low cloud											*	*
15 Pollution content	*											
16 Wind speed	*		*	*	*	*				*	*	*
17 Wind direction	*		*	*	*					*	*	*
18 Wind vertical variation										*	*	
19 Rain amount						*	*					
20 Rain intensity				*			*					
21 Rain duration							*					
22 Snow amount					*				*			
23 Evaporation												

Table 2 Meteorological information requirements for design.

| | Structural Stability | | | | | | Services | Weather tightness | | | | | | | |
| | 2/1 | 2/2 | 2/3 | 2/4 | 2/5 | 2/6 | 2/7 | 2/8 | 2/9 | 2/10 | 2/11 | 2/12 | 2/13 | 2/14 | 2/15 |
	Wind loading	Snow loading	Ice loading	Temperature extremes on materials	Foundations - moisture movements	Foundations - frost heave	Freezing of buried pipes	Rain penetration through cracks	Rain absorption by materials	Snow penetration through cracks	Air leakage through cracks	Natural ventilation	Draught in chimneys	Roof drainage	Rain run-off from walls
1 Solar radiance, direct on horizontal				✻	✻										
2 Solar radiance, diffuse on horizontal				✻	✻										
3 Illuminance direct on horizontal															
4 Illuminance diffuse on horizontal															
5 Sunshine duration															
6 Sunpaths															
7 UV radiance total on horizontal															
8 Equivalent radiative temperature of sky				✻											
9 Temperature of air at standard height		✻		✻	✻	✻	✻				✻				
10 Temperature at vertical variation				✻							✻				
11 Temperature in ground						✻	✻								
12 Humidity of air					✻										
13 Fog				✻											
14 Low cloud				✻											
15 Pollution content															
16 Wind speed	✻	✻	✻	✻	✻			✻	✻	✻	✻	✻	✻		✻
17 Wind direction	✻	✻	✻					✻	✻	✻		✻	✻		✻
18 Wind vertical variation	✻		✻								✻	✻	✻		
19 Rain amount		✻			✻				✻					✻	
20 Rain intensity								✻	✻					✻	✻
21 Rain duration															
22 Snow amount			✻	✻						✻					
23 Evaporation					✻										

Table 2 Meteorological information requirements for design (cont).

	Heating/cooling				Daylighting				
	2/16	2/17	2/18	2/19	2/20	2/21	2/22	2/23	2/24
	Heating	Cooling	Thermal insulation	Warm nights	Illuminance in rooms	Sky luminance distribution	Daylight quality – spectral distribution	Luminous efficacy	Sunshine in buildings
1 Solar radiance, direct on horizontal	✻	✻						✻	
2 Solar radiance, diffuse on horizontal	✻	✻							
3 Illuminance direct on horizontal					✻			✻	
4 Illuminance diffuse on horizontal					✻	✻	✻		
5 Sunshine duration					✻				✻
6 Sunpaths					✻				✻
7 UV radiance total on horizontal									
8 Equivalent radiative temperature of sky	✻	✻							
9 Temperature of air at standard height	✻	✻	✻	✻					
10 Temperature of air at vertical variation	✻	✻							
11 Temperature in ground									
12 Humidity of air		✻	✻						
13 Fog									
14 Low cloud									
15 Pollution content									
16 Wind speed	✻	✻		✻					
17 Wind direction				✻					
18 Wind vertical variation									
19 Rain amount									
20 Rain intensity									
21 Rain duration									
22 Snow amount									
23 Evaporation									

Table 3 Meteorological information requirements for construction process.

	Physiological effects				Site conditions					Effects on materials				
	3/1	3/2	3/3	3/4	3/5	3/6	3/7	3/8	3/9	3/10	3/11	3/12	3/13	3/14
	Amount of daylight available	Precipitation (rain, hail, snow) effect on workers	Thermal stress on workers	Risk of accidents	Flooding of works	State of gound (wetness)	State of ground (snow and frost)	Traffic on and off the site	Operation of cranes	Cold weather working	Hot weather working	Drying out of structure	Wetting of concrete	Plastic cracking of concrete
1 Solar radiance, direct on horizontal			*			*								
2 Solar radiance, diffuse on horizontal			*											
3 Illuminance direct on horizontal														
4 Illuminance diffuse on horizontal	*													
5 Sunshine duration														
6 Sunpaths	*													
7 UV radiance total on horizontal														
8 Equivalent radiative temperature of sky														
9 Temperature of air at standard height		*	*	*						*	*	*		
10 Temperature of air at vertical variation														
11 Temperature in ground							*			*				
12 Humidity of air			*	*								*		*
13 Fog								*						
14 Low cloud														
15 Pollution content														
16 Wind speed		*	*	*					*	*		*		*
17 Wind direction									*					
18 Wind vertical variation									*					
19 Rain amount					*							*	*	
20 Rain intensity		*		*	*									
21 Rain duration		*												
22 Snow amount							*	*						
23 Evaporation						*								

Table 4 Meteorological information requirements for maintenance and running costs.

	4/1 Frost action	4/2 Photoactinic action	4/3 Thermal movements	4/4 Moisture movements	4/5 Corrosion of metals	4/6 Condensation and mould growth	4/7 Heating costs	4/8 Cooling costs	4/9 Lighting costs	4/10 Dirtying of exteriors
1 Solar radiance, direct on horizontal			*				*	*		
2 Solar radiance, diffuse on horizontal			*				*	*		
3 Illuminance direct on horizontal									*	
4 Illuminance diffuse on horizontal									*	
5 Sunshine duration										
6 Sunpaths										
7 UV radiance total on horizontal		*								
8 Equivalent radiative temperature of sky							*	*		
9 Temperature of air at standard height	*		*	*		*	*	*		
10 Temperature of air at vertical variation										
11 Temperature in ground										
12 Humidity of air				*	*	*		*		
13 Fog										
14 Low cloud										
15 Pollution content					*					*
16 Wind speed			*	*		*	*	*		*
17 Wind direction										*
18 Wind vertical variation										*
19 Rain amount	*			*						
20 Rain intensity										*
21 Rain duration										
22 Snow amount										
23 Evaporation										*

The various types of solar radiation data that may be recorded or required are:

a) direct radiation at normal incidence,
b) direct and diffuse radiation at normal incidence,
c) direct radiation on a horizontal plane,
d) direct and diffuse radiation on a horizontal plane,
e) data on spectral distribution of direct, diffuse and total energy.

The form of the data most often recorded is daily sums of the total radiation incident on a horizontal surface. Monthly averages of these daily data are most frequently required. Measurement is usually by thermo-electric or bimetallic expansion pyranometers. Sunshine hours (as opposed to quantitative radiation data) are most frequently recorded by a Campbell-Stokes sunshine recorder.

Report No 21* of the College of Engineering, University of Wisconsin, gives overall data on the world distribution of solar radiation and gives extrapolated data for those areas where no radiation data are available and, in addition, gives an approximate formula for changing hours of sunshine into total radiation.

Mists, fog, snow, dust storms, etc
Obviously data on the frequency of occurrence, duration and intensity of these should be collected where appropriate.

Miscellaneous data
Good topographical data together with vegetation cover (type, height and location) are required for an accurate picture of a site's microclimate.

Sources

The sources for the above information are numerous. However, it is extremely difficult to obtain complete information. Information is easily obtained for the UK from the Meteorological Office, Bracknell. The Ove Arup Partnership (OAP) already have the Kew weather data for 1969, including radiation, on an hourly basis on tape. Weather tapes are also available for most areas of Saudi Arabia and the Gulf States.

Finally, a local farmer will probably understand more about how the terrain, etc, modifies the microclimate than any other person, since he will have learnt from experience where and when and what crops he can plant.

SOME FACTORS AFFECTING MICROCLIMATE

The factors affecting microclimate are numerous and only some of these will be outlined here. There is a need for more research and data in this field and this has been recognised by most researchers, eg, P O'Sullivan, " Heat Islands in Cities", ART, May 1970.

Topography

Temperature
Temperature in the atmosphere decreases with altitude at the rate of approximately 1°F per 330 ft in summer and 1°F per 400 ft in winter.

Cool air is heavier than warm air and at night the outgoing radiation causes a cold-air layer to form near the ground surface. This is particularly pronounced in hot arid climatic zones. Consequently, the cold air will flow to the lowest points causing cold ponds in depressions or very large cold ponds in valleys. Like water, the cold air can also be dammed by an obstruction.

Water can have a pronounced effect on microclimate since it has a higher specific heat than the land. Thus it is normally warmer in winter and cooler in summer, and usually cooler during the day and warmer at night than the surrounding terrain. For example, in the Great Lakes region this effect raises the average January temperatures by about 5°F, the absolute minimum temperatures being about 10°F and the annual minima about 15°F. There is a similar but reduced depressive effect on summer temperatures.

The quantity of solar radiation also has a dominant effect on climate. A hillside receives radiation depending on the inclination and direction of the slope. For example, a south-facing slope of 40°N latitude at 20° inclination will receive 50% more total radiation in December than an equivalent level site and 200% more radiation than a north-facing slope of 20° inclination.

---

*See also article by R H Collingbourne in Solar Energy Society UK Section, February 1973.

Wind

Topography can significantly influence wind velocities and directions. Published data are normally recorded at level sites with relatively undisturbed air flows, viz airports. The two standard UK references are " Code of Basic Data for the Design of Buildings", Chapter V "Loading", Part 2: "Wind Loads", BSI, 1970, and "Tables of Surface Wind Speed and Direction over the UK", HMSO, MET 0.792.

Approximate ideas on the modification of wind by topography can be obtained by consulting an expert such as T V Lawson of the Department of Aeronautical Engineering, the University of Bristol, or, if more detailed information is required, then this can be obtained by wind tunnel tests such as those used on the Plan Guinet Valescure project. These tests and the method of procedure have been described by K C Anthony in the Arup Journal, September 1974.

Precipitation

When moisture-bearing winds occur frequently from the same direction, then topography can have a significant effect on precipitation patterns. Where ground changes of more than 300 m occur, the windward side of the slope can expect to receive more rainfall than the seasonal average and the leeward side correspondingly less.

Vegetation

Temperature

Plant and grassy covers reduce temperatures by the absorption of insolation and cool by evaporation. It has been found that temperatures over grassy surfaces on sunny summer days are about 10–14°F cooler than those of exposed surfaces.

Wind

Wind velocities can be very effectively moderated by planted shelter belts, and this brings perceptible changes, both in the temperature and humidity of the air, in evaporative effects and in the formation of snow-drifts. References are:

1) Olgyay, "Design with Climate".
2) Woodruff, N P, "Shelter Belt and Surface Barrier Effects", Agricultural Experiment Station, Manhattan, Kansas Technical Bulletin 77, December 1954.
3) "Shelter Belts and Microclimate", Forestry Commission Bulletin 29.

The ideal wind break has a porosity of approximately 35% and will give a 50% reduction of wind velocity for a distance of about 10/13 times the height of trees and a 25% reduction for a distance of approximately 27 times the height of trees.

The urban environment

Temperature

Cities and man-made structures tend to elevate temperatures. Helmut Landsberg has measured the temperature distribution in Washington, DC, which varied 8°F within horizontal distances of a few miles. At night the differences in temperature were even larger; some suburban territories had temperatures 11°F lower than downtown. This phenomenon has been confirmed by others, notably, P O'Sullivan ("Heat Islands in Cities") and T J Chandler ("The Climate of London"). P O'Sullivan, in his observations of the temperature gradients in Newcastle-upon-Tyne, has shown a difference of 3.8°C (7°F) and he concludes that at dawn the air temperature gradient between the urban environment and the periphery is at a maximum. During the day, radiation causes the air temperature to rise, generally, but the air temperature within the urban area does not increase at the same rate as that of the rural environs, and therefore the temperature gradient is reduced. There appear to be three reasons for this differential rate of temperature increase:

1. There are differences between the thermal capacities and conductivities of the fabric of the urban area, as compared with the vegetation covered soils, and so the air above the latter responds more quickly.

2. Heat build-up within the urban area is retarded as warmer air near the ground mixes with cooler air above. The mixing is caused by turbulence over the serrated surface of the city.

3. Heat generated by the city is dispersed in the more unstable daytime conditions. This, in turn, causes thermals over the area, and during the afternoon the gradient reaches a minimum.

As evening approaches and solar radiation ceases, the temperature rapidly falls off in rural areas, once again because of the thermal properties of the vegetation covered soils. By dusk the temperature gradient reaches a maximum – the 'onset of night time cooling'. The evening reduction of air temperature in the urban area is retarded by the increase in heat production from domestic grates, industrial boilers, vehicles and lighting, by the release of heat stored in the fabric of the city and by the reflection of heat from the walls of tall buildings and pollution haze. The contribution of each of these factors depends on the reflectivity, temperature conductivity, and thermal capacity of the surfaces.

During the evening the urban areas slowly cool down and the temperature gradient is again reduced. Thus two distinct temperature/time curves (diurnal temperature variation curves) exist: one external to the urban area, with a lower minimum value and a rapid rate of change, sensitive to general weather conditions and time of day, and one within the urban area, with a higher minimum value, a maximum value dependent on urban density, and a higher mean daily value that is stable, to a large extent independent of daily weather fluctuations and time of day, and with a slow rate of change. The mean value changes primarily with the time of year.

Precipitation
A similar effect to that described under 'Topography' can be found on the leeward and windward sides of groups of high buildings.

Atmospheric pollution
Waste products from machinery, boilers, etc, vehicle exhausts, tend to reduce direct solar radiation, but increase diffuse radiation and provide a barrier to outgoing radiation.

For example, radiation data for Kew are not necessarily meaningful when used for Central London.

Relative humidity
Relative humidity in the centre of cities may be 5—10% lower than in outlying areas, due to the absence of vegetation and the higher air temperatures already described.

Wind
Wind velocities may be much lower than in open countryside and this is recognised by CP3, Chapter 5, through the use of factor S2. However, local enhancements of velocities and turbulent effects, particularly adjacent to high buildings, can be very significant. Wind tunnel tests of both the building and its surroundings can be essential in studying these effects.

PRACTICAL EXAMPLES OF CLIMATE MODIFICATION

Two projects in different climatic regions illustrate this principle. The first example is for the hot dry arid climate of Riyadh, Saudi Arabia. The project was a recreational park. The climate adjacent to the buildings (microclimate) was modified by building location and orientation, planting, adjustment of topography and incorporation of hard and soft landscaping as appropriate. There is little comfort benefit from air movement in climates with low humidity and, in addition, in Riyadh there is a significant problem with airborne dust. Air movement adjacent to the buildings was modified by the use of judiciously located tree shelterbelts. Reflected radiation gain onto a building can represent between 8 and 13% of the total heat gain and this component was reduced by using planting immediately adjacent to its external walling.

The following are albedos (per cent) of various surfaces for total surface radiation with diffuse reflection:

Fresh snow cover	75—95
Clean firm snow	50—65
Light sand dunes	30—60
Sandy soil	15—40
Concrete	15—25
Meadows and fields	12—30
Woods	5—20
Dark cultivated soil	7—10

Changes in level in and around the building were used to emphasise the well-known nocturnal 'cold pond' effect, which is significant in a desert area because of high nocturnal radiation losses from a flat terrain to the clear sky. This results in cooler surface temperatures and the formation of a cool air layer adjacent to the ground which will tend to flow slowly into and collect in depressions within the terrain.

Fountains and open water features were located adjacent to covered pedestrian routes. The evaporative cooling effects adjacent to these features help to produce cool, pleasant conditions.

Figure 2 Windfield pattern around buildings.

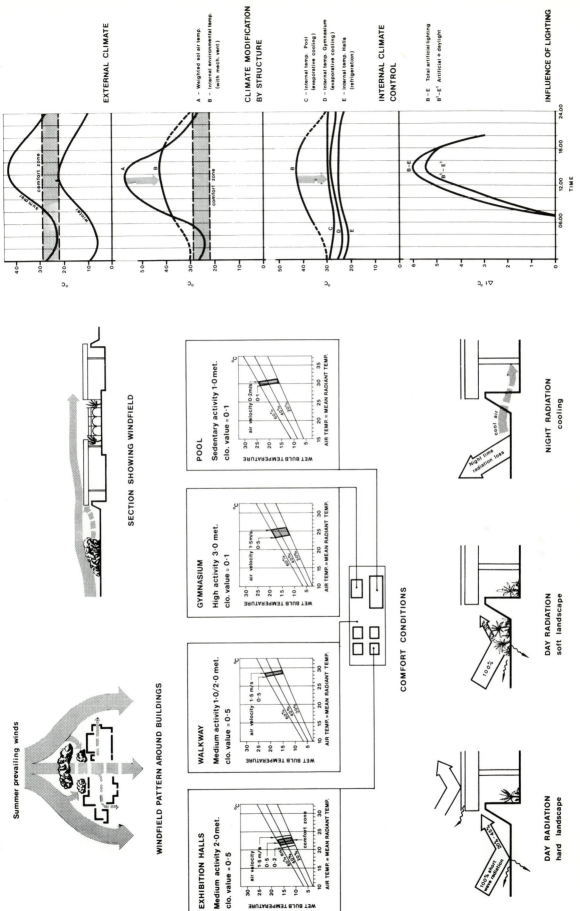

EXTERNAL CLIMATE

A – Weighted sol air temp.
B – Internal environmental temp. (with mech. vent)

CLIMATE MODIFICATION BY STRUCTURE

C – Internal temp. Pool (evaporative cooling)
D – Internal temp. Gymnasium (evaporative cooling)
E – Internal temp. Halls (refrigeration)

INTERNAL CLIMATE CONTROL

B – E   Total artificial lighting
B¹–E¹   Artificial + daylight

INFLUENCE OF LIGHTING

Summer prevailing winds

WINDFIELD PATTERN AROUND BUILDINGS

SECTION SHOWING WINDFIELD

EXHIBITION HALLS
Medium activity 2·0 met.
clo. value = 0·5

WALKWAY
Medium activity 1·0/2·0 met.
clo. value = 0·5

GYMNASIUM
High activity 3·0 met.
clo. value = 0·1

POOL
Sedentary activity 1·0 met.
clo. value = 0·1

COMFORT CONDITIONS

DAY RADIATION hard landscape

DAY RADIATION soft landscape

NiGHT RADIATION cooling

29

30

Figure 3 Windfield pattern.

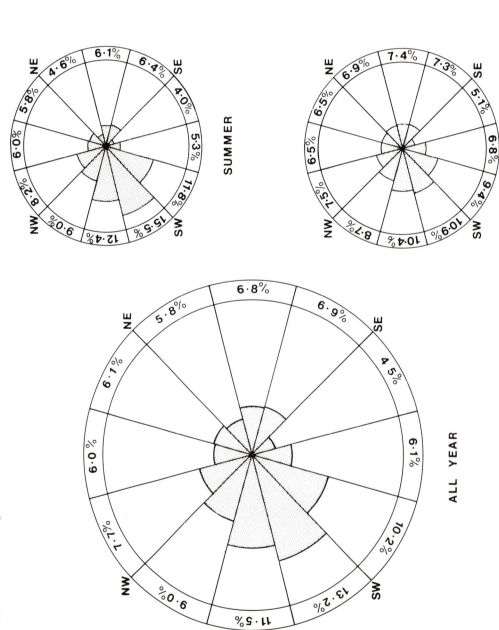

DESIGN DATA

1. Wind data based on readings recorded at CORYTON, ESSEX.

2. Period 1959 – 1964 compiled from hourly tabulations.

3. Results expressed as percentage number of hours over the year/season. Direction of wind shown as shaded portion of diagrams.

SUMMER

NE 6.1% 6.4% SE
4.6% 4.0%
5.8% 5.3%
6.0% 11.8%
8.2% 15.5%
9.0% 12.4%
NW SW

WINTER

NE 7.4% 7.3% SE
6.9% 5.1%
6.5% 6.8%
6.5% 9.4%
7.5% 10.9%
8.7% 10.4%
NW SW

ALL YEAR

NE 6.8% 6.9% SE
5.8% 4.5%
6.1% 6.1%
6.0% 10.2%
7.7% 13.2%
9.0% 11.5%
NW SW

DESIGN DATA

1. Sun's path plotted for latitude 51° – 29' North for months indicated.

2. Solar radiation data based upon IHVE Data Book 'A'.

3. Corrected solar radiation data used for all calculations provided by Meteorological Office, Kew, showing mean values of Global solar radiation for each hour of each month covering the ten year period 1959 – 1968.

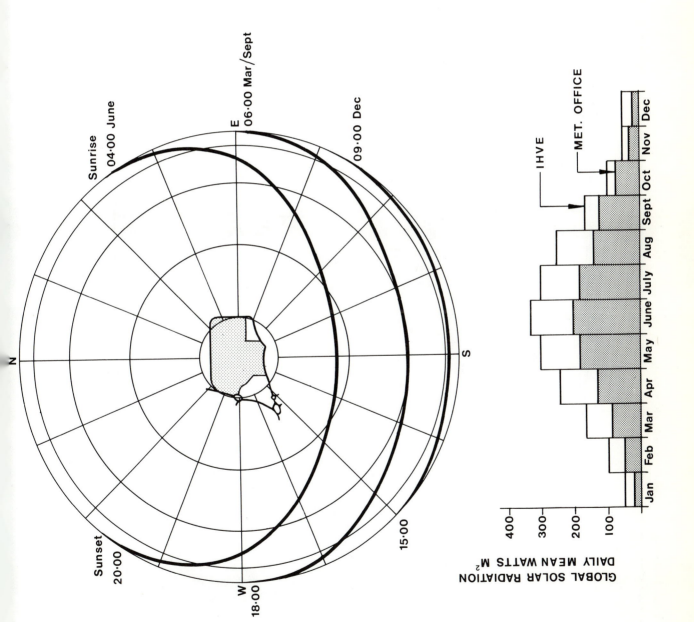

Figure 4 Sunpath.

Figure 5 Shelterbelt pattern and recommended spacing.

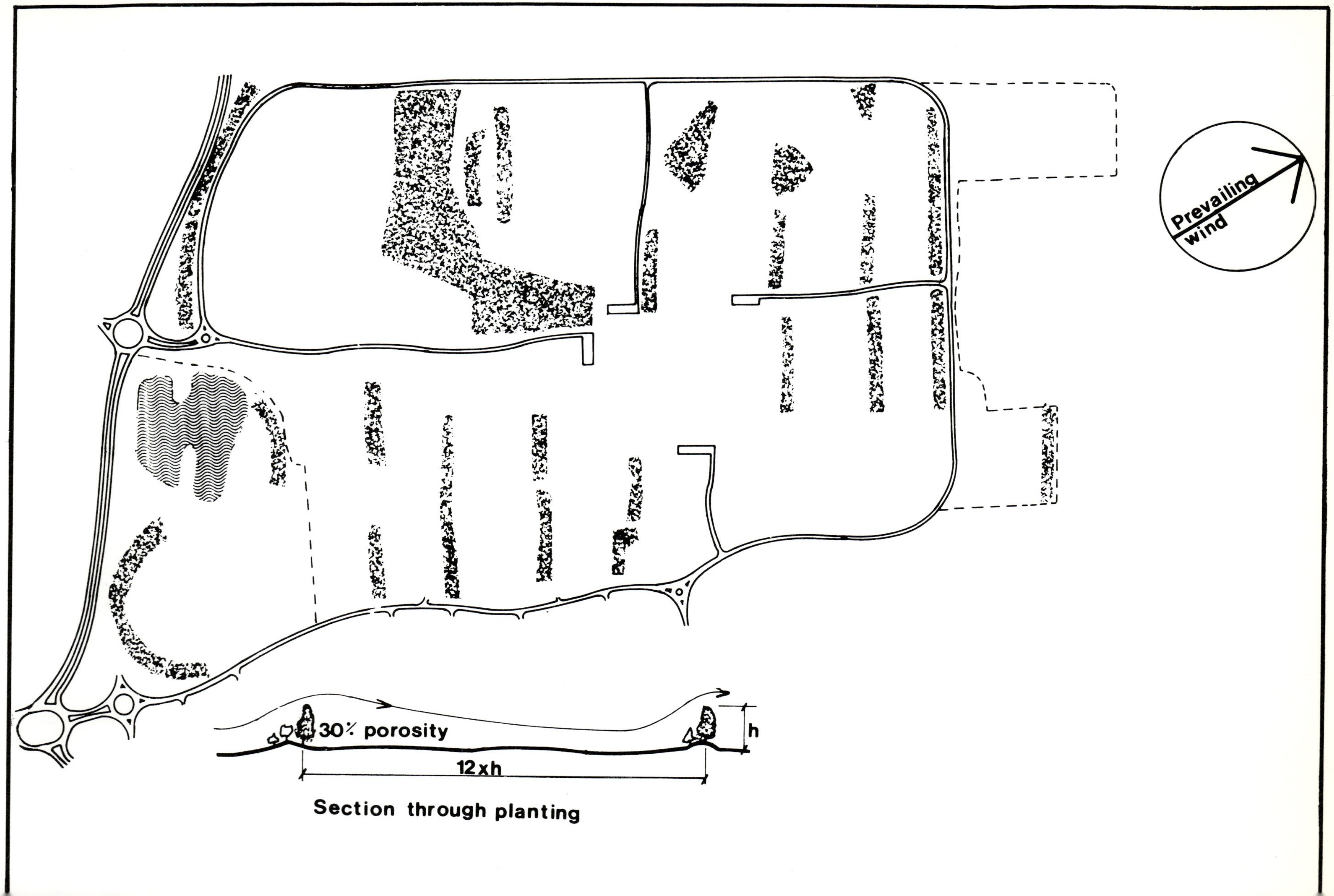

Figure 6 Increase in energy demand for the heating season due to overshadowing.

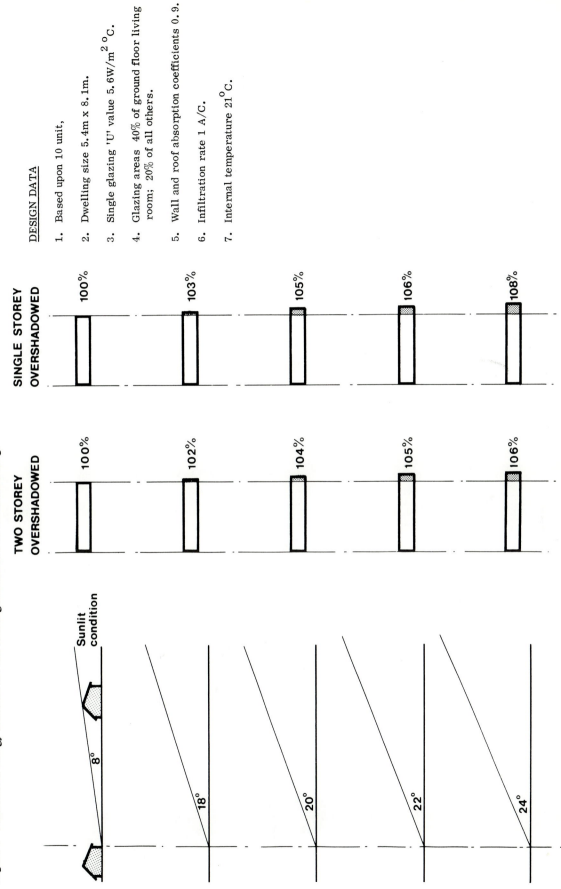

SINGLE STOREY OVERSHADOWED

100% 103% 105% 106% 108%

TWO STOREY OVERSHADOWED

100% 102% 104% 105% 106%

Sunlit condition

8°

18° 20° 22° 24°

DESIGN DATA

1. Based upon 10 unit,

2. Dwelling size 5.4m x 8.1m.

3. Single glazing 'U' value 5.6W/m^2 °C.

4. Glazing areas 40% of ground floor living room; 20% of all others.

5. Wall and roof absorption coefficients 0.9.

6. Infiltration rate 1 A/C.

7. Internal temperature 21°C.

33

Figure 7 **Northlands Housing, Basildon.** *(Photo: copyright John Donat)*

The building fabric itself was the final moderator of the harsh external climate. The major component of heat gain to the building is via the roof, and consequently a double skin, ventilated and insulated roof construction was adopted. The basic wall and roof construction was of high thermal inertia to dampen the diurnal variation in internal temperature (see Figure 2).

The second example is for the temperate climate of the UK. It is a low energy housing scheme in Basildon, Essex, where the peak heating demand of a 5–6 person house is in the range of 4–5 kW. For highly insulated housing it is vitally important to control air infiltration to the dwellings.

The prevailing wind directions on this site are WSW and ENE. Tree shelterbelts were planted normal to these wind directions to reduce site wind velocities and thus infiltration rates within the dwellings and the wind cooling of the fabric of the dwellings.

Free solar gain to the south-facing, terraced houses during the winter months makes a significant reduction in the demands on the installed heating system within the houses. Terrace spacing was determined to ensure that low angle winter sun could still enter the living spaces of the houses and overshadowing from adjacent terraces did not occur.

Figures 3–6 illustrate some of the studies carried out on this project jointly by the architect, the structural engineer and the services engineer.

THE NEXT STEP

Climates of enclosed spaces are referred to as cryptoclimates. The existence of an enclosure radically changes some climatic elements and hopefully excludes others, such as precipitation. The climatic elements of most interest within an enclosure are temperature, humidity and air movement. Correct design of the enclosure can significantly reduce the amplitude of the climatic variations and also alter the positions of the maximas and minimas by up to 12 hours on the diurnal range by the correct selection of walling and roofing materials.

The 'micro-cryptoclimate' is also attracting much interest. We might define this as the climate surrounding and particular to an individual. Schemes employing low levels of background illuminance and ventilation, and employing both lighting and ventilation local to the task and individual are already being designed and built.

ACKNOWLEDGEMENT

Figures 2.1–2.5 are taken from " Survey of Meteorological Information for Architecture and Building", Current Paper 5/72, Building Research Station, 1972, and are reproduced by permission of The Controller, HMSO (Crown Copyright).

CREDITS

Riyadh Recreational Park, Saudi Arabia.
Architect: The Architects Collaborative, Cambridge, Massachusetts, USA.

Felmore Housing, Basildon, Essex.
Architect. Ahrends, Burton and Koralek.
Client: Basildon Development Corporation.

BIBLIOGRAPHY

1   Koenigsberger, O H, Ingersoll, T G, Mayhew, A, and Szokolay, S V, "Manual of Tropical Housing and Building", Part I, "Climatic Design", Longman, 1974.
2   Yoshino, M M, "Climate in a Small Area", University of Tokyo Press, 1976.
3   Geiger, R, "The Climate Near the Ground", Harvard University Press, 1966.
4   Griffiths, J F, "Climate and the Environment", Elek, 1976.
5   Olgyay, "Design with Climate", Princeton University Press, 1963.
6   College of Engineering, University of Wisconsin, Report No 21.
7   O'Sullivan, P, "Heat Islands in Cities", ART, May 1970.
8   Lacy, R E, "Climate and Building in Britain", Building Research Establishment, 1977.

Chapter 7

Ventilation effects

M Holmes

INTRODUCTION

The energy used by heating and air-conditioning services in a building is expended in two ways:

1. Control of the dry-bulb temperature.
2. Control of the wet-bulb temperature (humidity control).

The degree of control – the amount of heat to be supplied or extracted – is dependent upon both internal loads and external conditions. The external loading comprises components related to temperature, intensity of solar radiation and wind speed. The relative importance of each of these will depend upon the design of the building and type and standard of construction. This chapter is concerned with the loads that arise from the combination of wind speed and external temperature – the infiltration load. (These parameters will also affect the thermal transmission through the fabric, which is not covered here; also only loads arising from dry-bulb temperature differences will be considered. Similar considerations can of course be applied to latent loads.)

MECHANISM OF INFILTRATION

Infiltration may be defined as the fortuitous leakage of air through a building due to imperfections in the structure. These imperfections may take the form of cracks around doors, windows, infill panels or between cladding sheets, etc. Building materials are porous so there will in general always be an infiltration of air from outside to inside.

A minimum of infiltration is of course desirable, for health reasons and prevention of condensation. We are concerned here with discussing the implications of infiltration loads beyond these minimum requirements on the assumption that the minimum levels have been achieved.

For air to pass through the openings there must be a driving force. This force is the pressure difference between the inside and outside of the building due to:

1. Wind.
2. Temperature difference.

Wind pressures require no explanation. The temperature difference effect, which is normally called the 'stack effect', is not always so easily understood.

The stack effect arises because air density varies with temperature. A building may be considered to contain a column of air (at the internal temperature), whilst outside there is a similar column at a different temperature. Clearly the two columns will have different weights, and thus give rise to a pressure difference across the building. For example, a 10 m high building 10°C above the outside temperature will have a maximum pressure difference across it due to stack effect of 4.3 Pa; this is equivalent to the dynamic pressure of a 2.7 m/s wind. The taller the building, the greater the stack effect (stack effect is proportional to height). Infiltration will therefore occur on both still and windy days.

HEAT LOAD DUE TO INFILTRATION

The sensible heat load arising from infiltration is obtained from the equation:

$$Q = m_a \, Cp\Delta t - (kW),$$

where

m_a is the air mass flow through the structure (kg/s),
C_p is the specific heat of air (kJ/kg °C),
Δt is the temperature difference across the building (°C).

It is, therefore, a very simple matter to calculate the energy consumed by excessive infiltration rates. The difficulty is to calculate the air flows involved. The MED note which forms the Appendix shows how infiltration can be calculated and also some data on the leakage characteristics of certain building elements. Table 1 in the note shows that as insulation standards improve infiltration comprises a greater proportion of the thermal load.

Example

The data given in Table 2 in the MED note does not show directly the relative significance of the different leakage characteristics of elements within the building fabric. This is probably best demonstrated by an example:

We will consider a simple rectangular building, with the following dimensions:

North and south elevations	:	25 m long x 7 m high,
East and west elevations	:	15 m long x 7 m high,
Roof	:	flat.

Glazing

North and south elevations	:	10 pivoted 1.5 x 1.5 m², sill height 1.5 m above ground level.
East elevation	:	5 windows similar to those on north and south elevation.
West elevation	:	2 windows similar to those on the north and south faces.

Doors
One standard external door on the east face, and a 5 m wide by 3.5 m high door on the west face.

The building is located on an industrial estate in SE England. Table 1 gives the calculated air change rates and corresponding heat loads, with a 4 m/s north wind, an external temperature of 7°C and an internal temperature of 20°C. Several constructional options are considered.

Option 1. Walls 216 mm plain brickwork. Roof asphalted concrete.

Option 2. Curtain wall. Roof as option 1.

Option 3. As option 2, but less well constructed, with an equivalent of a 10 mm gap around the perimeter of the building.

Table 1 Air change rates and heat loads.

Option	Airchange rate (per hour)	Heat load (kW)
1	0.14	1.65
2	0.22	2.55
3	0.48	5.63

Table 2 gives an indication of the significance of each element of building, by showing the percentage of the total air flow through each element of the fabric.

Table 2 Distribution of infiltration.

Elements of fabric	Inflow % for option			Outflow % for option		
	1	2	3	1	2	3
Wall	13	51	23	7	36	14
Window	35	20	8	0	1	2
Roof	0	0		93	63	28
Small door	10	6	3	0	0	
Large door	42	23	10	0	0	
10 mm crack			56			56

Note: Inflow figures are percentage of total infiltration into the building.
Outflow figures are percentage of total air leaving the building.

CONCLUSION

Infiltration will occur through all elements of a building structure. The importance of any single element of the fabric will depend upon the quality of the rest of the structure. Small gaps in the structure, for example badly fitted cladding panels, can significantly affect the rate of infiltration and consequently increase energy consumption. Just as important, a badly constructed building, with high infiltration rates, may prevent the heating system from achieving design conditions on cold days. The result is a dissatisfied client.

APPENDIX. MED NOTE 24 (September/October 1978), "Calculation of Natural Ventilation Rates" by M Holmes
(MED notes are a series of internal documents prepared by the Mechanical and Electrical Development Group and published as supplements to the Arup Newsletter.)

Introduction

Outside air can be supplied to a building by either natural or mechanical ventilation. In both cases supplying more air than necessay will result in an increase in the heat load on the building. Natural ventilation will occur due to leakage through the building fabric and purpose-made ventilators. If mechanical ventilation is provided there is still a very large probability that there will be infiltration through the building fabric as it is virtually impossible to construct an air-tight building. As building insulation standards improve, the significance of excessive infiltration increases. This is demonstrated by the figures given in Table 1 for a 3 x 3.5 x 5 m deep office with one external wall and an air change rate (infiltration) of half an air change per hour.

Table 1 Proportion of heat load due to infiltration.

External wall U value $(w/m^2\ {}^\circ C)$	Ratio of infiltration load to total heating load
5.6	0.13
2.5	0.25
1.0	0.46
0.5	0.63

It is clearly important that, as insulation standards rise, excessive infiltration is prevented. It is also necessary, for health purposes, that a minimum infiltration rate is always available. The object of this note is to outline the parameters affecting infiltration, and to discuss methods for calculating the leakage rate through the building fabric.

Parameters controlling natural ventilation

Natural ventilation rates are dependent upon:

a)   wind speed and direction,
b)   outside air temperature,
c)   internal air temperature,
d)   size and type of leakage paths through the building,
e)   location of the building.

For a given building, on a given site, the infiltration rate will, of course, vary only with wind velocity and the internal/external temperature difference. In theory it is necessary to calculate the infiltration rate using the combined effect of wind and temperature difference (note that the temperature difference effect is normally referred to as 'stack effect'). In practice this is not always necessary. Figure 2 gives the air change rate through the simple building shown in Figure 1 as a function of wind speed for an external temperature of $-1^\circ C$ and an internal temperature of $20^\circ C$. The solid line on Figure 2 shows the infiltration rate without stack effect, whilst the dotted line includes the combined effect of wind speed and stack. The slight reduction in infiltration rate at a wind speed of about 3 m/s is due to a change in the flow pattern through the building as the wind pressure overcomes the stack effect on the upstream face of the building. Figure 2 suggests that, for simple buildings, the infiltration rate may be approximated by calculating the rates for wind and stack separately and then using the larger of the two rates. Introducing a vent into the roof changes the infiltration rate curve to that shown in Figure 3. Again the infiltration rate can be approximated by using the larger of wind or stack effects. Wind and stack effect are *not* additive.

Figure 1 Section through simple building.

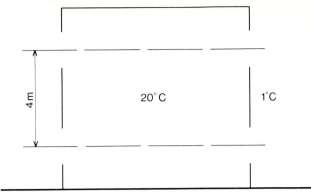

Figure 2 Air change rate for a simple building.

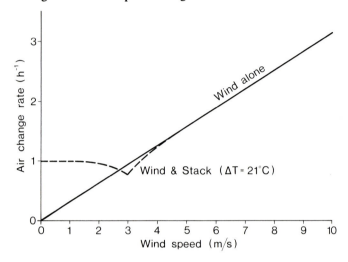

Figure 3 Air change rate for simple building with roof vent.

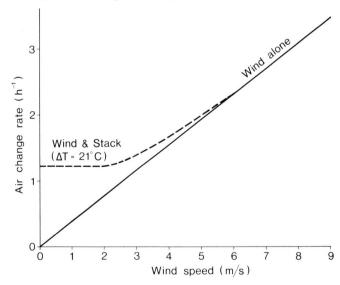

Calculation of infiltration rate for small openings

Ostensibly the calculation of infiltration is simple, although in general a computer program will be necessary to solve the equations.

Difficulties arise because of:

a) lack of knowledge of the characteristics of leakage paths through the building,
b) the pressure distribution, due to wind, on the surface of the building.

Leakage characteristics

The flow rate through a crack can be calculated from the equation:

$$Q = K (\Delta p)^N \qquad (1)$$

where

Q   is the flow rate (l/s),
Δp is the pressure difference across the crack (Pa),
K and N are constants.

NB. Both K and N will vary with the Reynolds number, but all the uncertainties involved in the calculations do not merit refinement of Equation (1).

Table 2 gives typical values of K and N for some elements of a building structure (1, 2).

Table 2   Air leakage data for building components.

Component	Description	K	N
Window (closed)	Pivoted	0.21/m crack	0.63
	Pivoted and weather-stripped	0.03/m crack	0.63
	Sliding	0.08/m crack	0.63
Doors (closed)	Single stairwell door	15/door	0.5
	Lift door	40/door	0.5
	External door with sill	1.6/m crack	0.5
	Standard fire stop door with		
	3 mm gap	2.6/m crack	0.5
	Lift door with 5 mm gap	4.2/m crack	0.5
		Per m²	
		surface	
Brick and masonry	8½ in (216 mm) plain brick	0.046	0.87
	8½ in (216 mm) with plaster	0.0004	0.87
	13 in (330 mm) plain brick	0.041	0.87
	13 in (330 mm) with plaster	0.0004	0.87
	External curtain wall	0.3	0.5
	Floors	0.2	0.5

The values for leakage through brick and masonry are for an average standard of construction and are therefore subject to variation from building to building.

Pressure distribution

Wind pressure. The only satisfactory method of obtaining the pressure distribution over the surface of a building is to use a wind tunnel test, where it is possible to model the building and surroundings. This is not always practicable and approximate pressures can be obtained using CP3: Chapter V (3). Wind tunnel tests have shown that if the values obtained from CP3 are multiplied by about 2/3 then the resulting pressures will be suitable for calculating infiltration rates (4).

Stack effect. The pressure difference between inside and outside of a building arising from temperature difference alone is calculated from Equation 2:

$$\Delta p = 3462\, h \left[\frac{1}{to + 273} - \frac{1}{ti + 273} \right] \qquad (2)$$

where

Δp is the pressure difference (Pa),
h   is the vertical distance between two openings (m),
to   is outside temperature (°C),
ti   is inside temperature (°C).

Combined wind and stack effect. The pressure difference resulting from the combination of wind and stack is the algebraic sum of the two pressure differences. In practice it is best to take a datum height within the building and calculate all stack pressures with respect to this level, for example:

If the pressure inside the building is Pi (at the datum level), the pressure on the surface due to wind is Pw, and if the crack is H above the datum then the pressure difference across the crack (ΔPc) is:

$$\Delta Pc = Pw - 3462\ H \left[\frac{1}{to + 273} - \frac{1}{ti + 273} \right] - Pi \qquad (3)$$

and the flow rate (Qc):

$$Qc = K\ (Pc)^N \qquad (4)$$

The infiltration rate through the building is obtained by solving a set of equations of the form of Equation 4.

Example

The general method of solution is to write the equation for flow rate through each crack and the continuity equation (flow into building is equal to flow out of building) and solve simultaneously. For example, taking the very simple building in Figure 4, with well-sealed walls:

Details:
Window: Type — pivoted dimensions 3 m x 1 m.
Door: Type — external with sill dimensions 2 m x 0.75 m.
Windspeed: 4 m/s.
Internal temperature: 20°C.
External temperature: −1°C.
Calculation of infiltration due to stack effect: Take reference height as centre of the door.
Pressure drop across door = Po − Pi, where Po is external static pressure and Pi is internal static pressure.

Figure 4 Elevation of simple building.

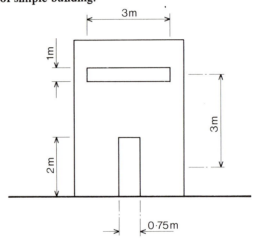

Flow rate through door $\qquad Q_D = K\ (Po - Pi)^N \quad$ (l/s) $\qquad (5)$

From Table 1, K = perimeter of door x 1.6,
$\qquad\qquad$ = 8.8,
$\qquad$ N = 0.5.

Hence $\qquad\qquad\qquad Q_D = 8.8\ (Po - Pi)^{0.5} \qquad (5a)$

Pressure drop across the window; assuming the flow is out through the window, Equation 3 becomes:

$$\Delta Pw = Pi - Po + 3462\ H \left[\frac{1}{to + 273} - \frac{1}{ti + 273} \right] \qquad (6)$$

with H equal to 3 m:

$$Pw = Pi - Po + 2.74.$$

Thus the flow rate through the window (Qw) is:

$$Qw = K\ (Pi - Po + 2.74)^N \quad \text{(l/s)} \qquad (6a)$$

where from Table 2:

K = window perimeter x 0.21,
$\qquad$ = 1.68,
N = 0.63,

42

ie, $Q_w = 1.68 (P_i - P_o + 2.74)^{0.63}$ (6b)

Now the air mass flow rate through the door must equal that through the window. It is however sufficiently accurate to set the volume flow rates equal thus:

$$Q_D = Q_w$$

or equating (5a) and (6b)

$$8.8 (P_o - P_i)^{0.5} = 1.68 (P_i - P_o + 2.74)^{0.63}.$$ (7)

If the only wind pressure information available is from CP3, then the pressure will have to be assumed uniform over the building surface and P_o (in this case) can therefore be set (conveniently) to zero. Thus it is only necessary to solve:

$$8.8 (-P_i)^{0.5} = 1.68 (P_i + 2.74)^{0.63}.$$ (7a)

Clearly P_i must be -ve. An iterative (or graphical) solution is required. The Newton-Raphson method is quite convenient, and in three steps yields a value for P_i of -0.122 (Pa).

Substituting into (5a) $Q_D = 3.08$ l/s

and check: Substituting into (6b) $Q_w = 3.08$ l/s. Thus the infiltration rate is 3.08 l/s.

The iterative solution results from the different values of N for each leakage path. It is clear that, as the building is made more complex, manual calculations will be very tedious.

Computer programs

An Arup program is now available. This calculates the ventilation rates for a building given:

1) wind speed, internal and external temperatures,
2) details of leakage paths and surface pressure coefficients.

Up to 20 rooms can be handled at present, with a maximum of 10 leakage paths to the outside; each room may be connected to any other by up to 10 parallel leakage paths. Any room may be provided with a fixed air supply or extract, and the effect of large openings with both inflow and outflow is covered.

The program calculates air change rates and heating or cooling loads due to the combined effect of wind, temperature difference and mechanical ventilation.

Where very simple buildings are involved (a simple building in this context is essentially a one-room building) the Newton-Raphson method should prove satisfactory, that is:

1) Set P_i to zero.

2) Calculate $\sum_{q=1}^{q=R} K_q (P_q - P_i)^{N_q} = f (P_i)$.

3) Calculate $\sum_{q=1}^{q=R} - K_q^{N_q} (P_q - P_i)^{(N_q - 1)} = f' (P_i)$.

4) New value of P_i = old value of $P_i - \dfrac{f(P_i)}{f'(P_i)}$.

5) Repeat until difference between old and new values is small.

6) Calculate flows for each crack, ie, $Q_q = K_q (P_q - P_i)^{N_q}$,

where

    q is the crack number,
    R is the total number of cracks,
    Q, P, K, N are as previously defined.

Large openings
An opening may be considered large if both inflow and outflow take place through the same orifice, examples would be open doors and windows.

Combined wind and temperature difference
If the external pressure (wind pressure) over the opening is uniform then the flow rate through the opening, due to wind pressure and temperature difference, can be calculated from the equation:

$$Q = \frac{2 \sqrt{2} C_D W}{3g (\rho_o - \rho_i) \sqrt{\bar{\rho}}} [(P_o - P_i) + g (\rho_o - \rho_i) H]^{3/2}$$ (8)

43

where

Q is the volume flow rate at mean density $\bar{\rho} \, \mathrm{m^3/s}$,
C_D is the discharge coefficient (normally about 0.6),
W is the width of the opening (m),
H is the height of the opening (m),
g acceleration due to gravity $(\mathrm{m/s^2})$,
ρ_o density of outside air $(\mathrm{kg/m^3})$,
ρ_i density of inside air $(\mathrm{kg/m^3})$,
$\bar{\rho}$ mean air density $= (\rho_o + \rho_i)/2$ $(\mathrm{kg/m^3})$,
P_o external wind pressure (Pa),
P_i internal pressure (Pa).

Table 3 gives calculated air flow rates for a range of opening heights, for a single opening into an otherwise sealed building. The external temperature is -1°C.

Table 3   Flow through a large opening.

Height of opening (m)	Volume flow rate/m width $(m^3/s$ per m) internal temperature	
	16°C	20°C
1	0.15	0.17
2	0.44	0.48
3	0.80	0.89
4	1.23	1.37
5	1.72	1.91
7	2.86	3.16
10	4.88	5.40

If there are other leakage paths then Equation (8) is solved simultaneously with the equations for flow through these.

Where the opening is a window, or door, the discharge coefficient will vary with the angle of opening. The Building Research Establishment have put forward the following factors to take account of this effect. They are given in Table 4.

Table 4   Discharge coefficient correction factors.

Angle of opening	Multiply the C_D by
0	0
15	0.4
30	0.65
45	0.80
60	0.90
90	1.00

Note. The above are averages of values for various window types and sizes given in Reference 1.

Wind alone
The air flow rate through a large opening into an otherwise sealed space due to wind pressure only can be calculated from:

$$Q = 0.025 \, A \times \text{reference wind speed} \ (\mathrm{m^3/s}) \qquad (9)$$

where A is the opening area $(\mathrm{m^2})$ and the reference wind speed (m/s) is the wind speed measured at a height equal to that of the building in free wind.

REFERENCES

1 Building Research Establishment, Digest 210, "Principles of Natural Ventilation", BRE, 1978.
2 Potter, I N, BSRIA, private communication.
3 British Standards Institution, CP3: Chapter V: Part 2: "1972 Code of Basic Data for the Design of Buildings". Chapter V, "Loading", Part 2, "Wind Loads", BSI, 1972.
4 Potter, I N, BSRIA, private communication.

Chapter 8

Numeracy aids for the assessment of environmental effects

J Campbell

INTRODUCTION

The first genuine numeracy aid, as opposed to a counting stone, was the abacus, which continued almost unchanged for hundreds of years. The rate of development since the industrial revolution has, however, forcefed development in many fields and this area is no exception. Log tables, mechanical calculators and slide rules have all been used extensively by engineers to assist them in their work. Now there are the electronic calculator, desk top computers, mini-computers and large mainframe computers.

These electronic numeracy aids have become more and more common over the last few years and have also reached a high level of sophistication. The hand calculator is now common-place. Desktop computers are readily available over the counter in stores and even large mainframe computers are easily accessible to anyone who wishes to use them via the medium of bureaux.

Numeracy aids normally fulfil one or more of main functions:

1. They reduce the time taken to carry out a series of calculations, thereby making the designer more cost effective.

2. They enable very complex calculations to be carried out with a degree of accuracy which would not have been feasible before their development.

3. They can help by producing the calculations in a standard form which simplifies checking and makes it easier for a job to be transferred from one engineer to another.

At a simple level the assistance is not given in the form of a numeracy aid, but rather in the simplification of its use. This takes the form of a calculation sheet for use with a hand calculator, and is a logical progression from slide rule calculation. At a more sophisticated level the numeracy aid is in the form of a computer program and this has the advantage that, once it has been thoroughly debugged, given the right input data it will produce the right answers by a known and agreed calculation method.

THE INTERNAL ENVIRONMENT

When a person enters a building he or she is subjected to several environmental effects which affect, to some extent, both the comfort and performance of that individual. The main factors involved in assessing this internal environment are:

a) thermal,
b) ventilation,
c) illumination,
d) sound and vibration.

The building itself, however, is also a climatic modifier in that it is designed to reduce or minimise — hopefully — the external environmental effects. Building services calculations are therefore carried out at two levels:

1. An assessment of the building form and the consequent internal conditions to determine the level of comfort to be — or that which will be — provided. This can be compared with the design requirements and a decision made about the type of system which must be provided.

Figure 1 Calculation sheet.

OVE ARUP PARTNERSHIP
Internal Temperature Estimate Sheet

Job Nº   Sheet Nº

Calc. by   Date

Room Reference		Classification		External Surface Colour	
Exposure of Room		Altitude of Room (m)			
Month of Peak Solar Intensity		Local Time of Peak Solar Intensity (h)			
Peak Solar Intensity, Glass 1 (W/m^2)		Mean Solar Intensity, Glass 1 (W/m^2)			
Peak Solar Intensity, Glass 2 (W/m^2)		Mean Solar Intensity, Glass 2 (W/m^2)			
Peak Solar Intensity, Glass 3 (W/m^2)		Mean Solar Intensity, Glass 3 (W/m^2)			
Solar Gain Factor, Glass 1		Solar Gain Factor, Glass 2		Solar Gain Factor, Glass 3	
Alt.Sol. Gain Factor, Glass 1		Alt. Sol. Gain Factor, Glass 2		Alt. Sol. Gain Factor, Glass 3	
Nett Area, Glazing 1 (m^2)		Admittance Value, Glass 1 $(W/m^2\ ^oC)$			
Nett Area, External Wall 1 (m^2)		Admittance Value, External Wall 1 $(W/m^2\ ^oC)$			
Nett Area, Glazing 2 (m^2)		Admittance Value, Glass 2 $(W/m^2\ ^oC)$			
Nett Area, External Wall 2 (m^2)		Admittance Value, External Wall 2 $(W/m^2\ ^oC)$			
Nett Area, Glazing 3 (m^2)		Admittance Value, Glass 3 $(W/m^2\ ^oC)$			
Nett Area, Roof or Ceiling (m^2)		Admittance Value, Roof or Ceiling $(W/m^2\ ^oC)$			
Nett Area, Floor (m^2)		Admittance Value, Floor $(W/m^2\ ^oC)$			
Nett Area, Internal Wall (m^2)		Admittance Value, Internal Wall $(W/m^2\ ^oC)$			
U Value, Glass 1 $(W/m^2\ ^oC)$		U Value, Glass 2 $(W/m^2\ ^oC)$		U Value, Glass 3 $(W/m^2\ ^oC)$	
U Value, External Wall 1 $(W/m^2\ ^oC)$		U Value, External Wall 2 $(W/m^2\ ^oC)$			
U Value, Roof or Ceiling $(W/m^2\ ^oC)$		U Value, Floor $(W/m^2 .^oC)$			
Density, External Wall 1 (kg/m^3)		Density, External Wall 2 (kg/m^3)			
Time Lag, External Wall 1 (h)		Time Lag, External Wall 2 (h)			
Decrement Factor, External Wall 1		Decrement Factor, External Wall 2			
Density of Roof (kg/m^3)		Density of Floor (kg/m^3)			
Time Lag, Roof (h)		Time Lag, Floor (h)			
Decrement Factor, Roof		Decrement Factor, Floor			
Length of Room (m)		Width of Room (m)			
Height of Room (m)		Volume of Room (m^3)			
Sol-air Temperature, External Wall 1 (^oC)		Mean Sol-air Temperature, External Wall 1 (^oC)			
Sol-air Temperature, External Wall 2 (^oC)		Mean Sol-air Temperature, External Wall 2 (^oC)			
Sol-air Temperature, Roof (^oC)		Mean Sol-air Temperature, Roof (^oC)			
Outside Air Temperature (^oC)		Mean Outside Air Temperature (^oC)			
Air Change Rate (h^{-1})		Ventilation Allowance $(W/m^3\ ^oC)$			
Total Internal Surface Area (m^2)		Ventilation Loss $(W/m^3\ ^oC)$			
Number of Occupants		Heat Emission/Occupant (W)		Duration of Occupancy (h)	
Number of Light Fittings		Heat Emission/Light Fitting (W)		Duration of Illumination (h)	
Equipment					

Mean Solar Heat Gain, Q'$_s$ =(x   x   )+(x   x   )+(x   x   )

Mean Casual Heat Gain, Q'$_c$ = (x   x   )+(x   x   )+(x   x   ) / 24

Total Mean Heat Gain, Q'$_t$ =   +

Exposed Areas x U Values, ΣAU = (x   )+(x   )+(x   )+(x   )

Ventilation Loss, C_v $\approx$ 0.33 x   x   : or $1/C_v$ = $1/0.33$ x   x   +1/4.8(+   )

Mean Internal Environmental Temperature = [(+   ) $(t'_{ei} -$)+($(t'_{ei} -$)]

Solar Swing, $\tilde{Q}_s$ = [x (-   )]+[x (-   )]+[(x (-   )]

Fabric Swing, $\tilde{Q}_f$ = [x   x (-   )]+[x   x (-   )]+[(x   x (-   )]

Casual Swing $\tilde{Q}_c$ = [(x   )+(x   )+(x   )+(x   )] -

Air to Air Swing, $\tilde{Q}_a$ = [(x   )+(x   )+(x   )+   ]x

Total Swing, $\tilde{Q}_t$ =   +   +   +

Value of ΣAY = (x   )+(x   )+(x   )+(x   )+(x   )+(x   )+(x   )

Swing in Internal Environmental Temperature = [(+   )] $\tilde{t}_{ei}$

Peak Internal Environmental Temperature, t''_{ei} =   +

MED 23 3/74

46

Figure 2 Amortisation of capital costs – graphical solution.

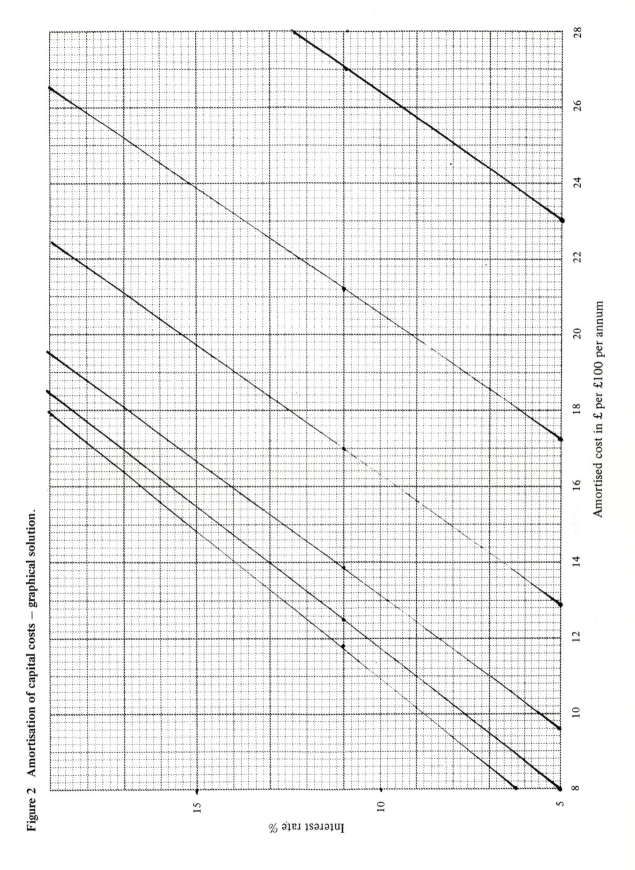

Interest rate %

Amortised cost in £ per £100 per annum

2. To determine the effect that the external environment will have on the equipment installed in the building and thus determine both its capacity and the cost of operating it.

Thermal environment

Although man can keep himself at the correct body temperature by involuntary physiological mechanisms over a wide range of external climates, he is comfortable in the real sense only over a comparatively narrow range of temperatures.

Under normal conditions of light work, the basic physiological process liberates about 110 W. If a normal design condition is taken of about 22°C, then about 80% of this is lost from the body by convection and radiation and 20% by evaporation, mainly from the lungs. If the work rate is increased, with a corresponding increase in the metabolic rate, then more heat must be lost from the body and the latent heat emission rises very rapidly.

Many different criteria have been developed over the years for judging comfort and that which currently meets with the most general acceptance is Dry Resultant Temperature. This scale was devised by Missenard in 1935. It takes into account air temperature, radiation and air speed. The dry resultant temperature is the temperature recorded by a thermometer at the centre of a blackened globe 100 mm in diameter and, as a consequence of the smaller globe size, it is rather less sensitive to radiation than the globe thermometer. It is related to the air and mean radiant temperature by the equation:

$$t_{res} = \frac{t_r + 3.17\, t_a \sqrt{v}}{1 + 3.17 \sqrt{v}}$$

where t_{res} = dry resultant temperature (°C).

Another criterion which has received acceptance over the past ten years is that of environmental temperature. It is in fact nothing new as it is a hypothetical temperature which enables a combined radiation and convection surface coefficient to be used.

At reasonable temperatures and velocities of air movement, the environmental and dry resultant temperatures are quite close together, meaning that the temperature used in calculations can be the same as that calculated for comfort purposes.

Ventilation

Ventilation has three main functions:

1. To provide a continuous supply of fresh air for breathing.
2. To remove products of respiration.
3. To remove 'artificial' contaminants arising from cooking or industrial processes.

In general, all these functions are discharged at the same time so that calculation of the fresh air supply should be based upon that function which requires the most air.

Ventilation can take two forms, controlled and infiltration. The latter is the energy-wasteful item and was dealt with in detail in the previous chapter. The former is essential for health reasons, but if it is introduced in a way which gives rise to excessive air movement it can affect the dry resultant temperature and therefore the level of thermal comfort.

Illumination

The eye has developed to be responsive to the wavelengths of radiation in which the sun gives off most of its energy. The human eye is a very adaptable organ. It is possible to read a newspaper by moonlight with an illumination level of 0.5 lux, whereas at the other extreme sunny days can produce illumination levels in excess of 25 000 lux.

The illumination level required in a space depends on the visual characteristics of the task. The most important of these are the size of the smallest detail and the contrasts and reflectances which occur within the task and its immediate surroundings. Typical illumination levels are given in Table 1.

Table 1.

Visual task	Illumination level (lx)
Casual	100
Rough tasks with large detail	200
Ordinary tasks with medium-size detail	400
Fairly severe tasks with small detail	600
Severe prolonged tasks with small detail	900
Very severe prolonged tasks with very small detail	1300–2000
Exceptionally severe tasks with minute detail	Over 2000

Table 2 Equivalent temperature differences for dark coloured internally insulated walls.

AUGUST

EXPOSURE	WALL THICKNESS M.M.	7 AM	8 AM	9 AM	10 AM	11 AM	12 AM	1 PM	2 PM	3 PM	4 PM	5 PM	6 PM	7 PM	8 PM	9 PM	10 PM
										SUN TIME							
NORTH	50	-4.6	-4.3	-2.1	0.0	2.0	3.5	4.9	5.5	5.9	5.4	4.4	3.1	3.0	-0.9	-2.4	-3.8
	150	-5.7	-5.8	-5.7	-5.3	-3.3	-3.1	-2.0	-0.8	0.2	1.0	1.8	2.1	2.3	2.1	1.5	0.8
	250	-3.3	-3.5	-3.6	-3.7	-3.6	-3.4	-2.5	-2.4	-1.9	-1.3	-0.8	-0.4	-0.1	0.0	0.1	0.0
	350	-2.1	-2.2	-2.4	-2.6	-2.7	-2.7	-2.8	-2.7	-2.6	-2.2	-2.1	-1.8	-1.5	-1.3	-1.1	-0.9
NORTH-EAST	50	4.9	10.1	9.8	6.1	2.2	3.7	5.0	5.7	6.0	5.6	4.6	3.2	1.4	-0.9	-2.2	-3.6
	150	-4.8	-4.9	-4.8	-4.2	2.6	5.4	5.2	3.3	1.1	2.0	2.7	3.0	3.2	3.0	2.5	1.7
	250	-1.9	-2.1	-2.2	-2.3	-2.2	-1.9	1.2	2.5	2.4	1.5	0.5	0.9	1.2	1.4	1.5	1.4
	350	-0.4	-0.6	-0.8	-1.0	-1.1	-1.1	-1.2	-1.1	-1.0	0.6	1.3	1.2	0.8	0.2	0.4	0.6
EAST	50	7.8	17.7	22.1	21.7	18.2	12.4	5.3	6.0	6.3	5.9	4.9	3.5	1.6	-0.7	-2.0	-3.4
	150	-3.3	-3.4	-3.3	-2.7	5.5	10.8	13.2	13.0	11.1	7.9	4.1	4.5	4.7	4.4	3.9	3.2
	250	0.2	0.0	-0.0	-0.1	-0.0	0.1	4.0	6.5	7.7	7.5	6.7	5.2	3.4	3.6	3.7	3.5
	350	1.9	1.8	1.6	1.4	1.3	1.2	1.2	1.2	1.4	3.4	4.7	5.3	5.3	4.8	4.0	3.1
SOUTH-EAST	50	2.6	12.8	20.2	24.7	25.9	24.6	20.6	14.7	7.5	6.0	5.0	3.6	1.7	-0.5	-1.8	-3.2
	150	-2.5	-2.6	-2.5	-2.0	3.4	8.9	12.9	15.3	16.0	15.2	13.1	9.9	6.0	5.2	4.7	4.0
	250	1.4	1.2	1.1	1.0	1.1	1.3	3.9	6.5	8.4	9.5	9.8	9.5	8.5	7.0	5.1	4.8
	350	3.3	3.1	3.0	2.8	2.7	2.6	2.6	2.6	2.8	4.1	5.5	6.4	7.0	7.2	7.0	6.5
SOUTH	50	-5.9	-2.1	5.5	13.2	19.8	24.6	26.7	26.6	23.6	18.7	12.2	5.3	1.7	-0.5	-1.8	-3.2
	150	-2.5	-2.6	-2.5	-2.3	-1.1	0.8	4.9	9.1	12.6	15.2	16.4	16.3	14.7	12.1	8.5	4.8
	250	1.4	1.2	1.0	1.0	1.0	1.2	1.7	2.7	4.6	6.6	8.2	9.4	10.0	9.9	9.2	8.0
	350	3.3	3.1	2.9	2.8	2.7	2.6	2.6	2.6	2.7	3.0	3.5	4.5	5.5	6.4	7.0	7.3
SOUTH-WEST	50	-5.9	-3.7	-1.6	0.6	3.6	12.7	20.6	26.6	29.8	30.2	26.9	20.2	10.3	-0.1	-1.8	-3.2
	150	-2.5	-2.6	-2.5	-2.2	-1.1	0.0	1.1	2.3	3.9	8.8	13.1	16.3	18.0	18.3	16.5	12.9
	250	1.4	1.2	1.1	1.0	1.1	1.2	1.7	2.3	2.8	3.4	4.2	6.5	8.5	10.0	10.8	10.9
	350	3.3	3.1	3.0	2.8	2.7	2.6	2.6	2.6	2.7	3.0	3.3	3.6	3.8	4.2	5.4	6.5
WEST	50	-6.0	-3.9	-1.7	0.4	2.4	4.0	5.3	14.3	22.1	27.2	28.8	25.2	15.5	-0.0	-2.0	-3.4
	150	-3.3	-3.4	-3.3	-3.0	-1.9	-0.8	0.3	1.5	2.6	3.4	4.1	9.0	13.2	15.9	16.8	14.8
	250	0.2	0.0	-0.0	-0.1	-0.0	0.0	0.5	1.1	1.6	2.2	2.7	3.1	3.4	5.7	7.6	8.9
	350	1.9	1.8	1.6	1.4	1.3	1.2	1.2	1.2	1.3	1.6	1.9	2.2	2.5	2.7	2.9	3.1
NORTH-WEST	50	-6.3	-4.1	-2.0	0.2	2.2	3.7	5.0	5.7	6.0	11.6	16.5	17.6	12.6	-0.3	-2.2	-3.6
	150	-4.8	-4.9	-4.8	-4.5	-3.4	-2.2	-1.0	0.1	1.1	2.0	2.7	3.0	3.2	6.2	8.8	9.4
	250	-1.9	-2.1	-2.2	-2.3	-2.2	-2.1	-1.6	-1.0	-0.5	0.0	0.5	0.9	1.2	1.4	1.5	2.9
	350	-0.4	-0.6	-0.8	-1.0	-1.1	-1.1	-1.2	-1.1	-1.1	-0.8	-0.5	-0.2	0.0	0.2	-0.4	0.6

EXPOSURE	WALL THICKNESS M.M.	7 AM	8 AM	9 AM	10 AM	11 AM	12 AM	1 PM	2 PM	3 PM	4 PM	5 PM	6 PM	7 PM	8 PM	9 PM	10 PM
											SUN TIME							

A more detailed classification is given in the IES code.

Illumination can be provided in two ways, naturally and artificially. Most rooms are provided with an adequate level of artificial lighting even when they are designed for natural illumination, as periods of occupancy normally exceed those in which natural daylight is available. In designing a room for daylight it is normal to calculate the daylight factor. This is the percentage of the external illumination level which reaches the desired point in the room.

Sound

The ear responds to frequencies in the range 15–20 000 Hz. The precise range differs from person to person. The ability to hear high-frequency sound steadily falls off with age due to deterioration in the nerves and muscles operating the hearing system.

Sound unwanted by the recipient is termed noise. As to all the other environmental factors, people react to sound in different ways. Care must be taken to protect people

Figure 3 Typical output for the daylighting program.

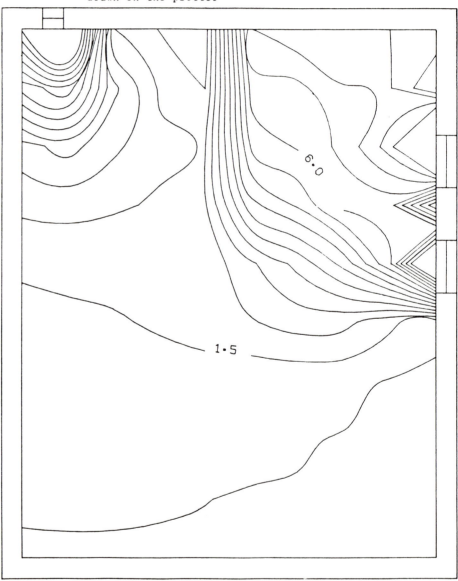

The Room Plan And Daylight Contours
drawn on the plotter

OAP/DOC DAYLIGHT ANALYSIS PROGRAM – ILLDH

DAYLIGHT FACTOR CONTOURS

FILE TITLE ROOM1·TST

from the nervous tensions and the physiological damage which can be caused by noise; the acoustical environment must be designed so that it is not detrimental to human performance.

One of the best practical ways of categorising noise is the noise-rating curve. Noise-rating curves are a method of rating the background noise level for annoyance and speech intelligibility in a given environment, and are an attempt to express equal human tolerance in each frequency band.

Noise can affect a room environment in three ways:

1. The internal room acoustics can be such that excessive sound pressure levels are generated.
2. External noise can penetrate the building fabric.
3. The mechanical services installed in the building may not only generate noise but also produce a route by which it can travel round the building.

DESIGN AIDS

Various calculation procedures have been produced to assist the designer in assessing these parameters, but in practice these can be quite involved and time consuming. There are various ways in which design aids can reduce the time taken in these calculations.

Figure 4 Graphical output from cooling load program (north-facing room in Northern hemisphere).

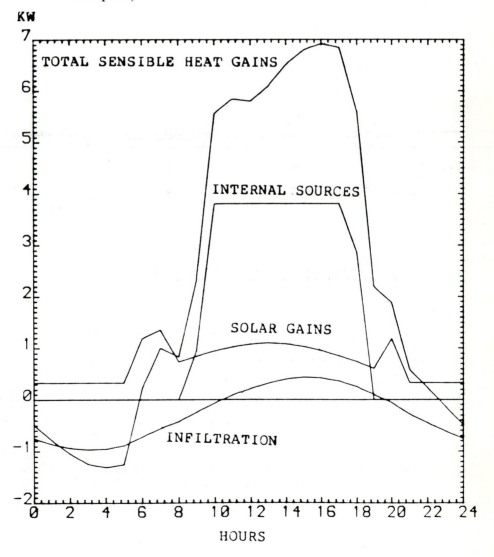

Figure 5 Typical input sheet for the energy program.

OVE ARUP PARTNERSHIP

COMPUTER DATA SHEET
THERMAL   sheet 1

Job no.   Sheet no. /
Filename
Made by
Date

**ENERGY PROGRAM - GENERAL DATA

JOB NUMBER >

JOB TITLE >

WEATHER FILE >   .   see manual for a list of available weather files

ALTITUDE <M> >

WEIGHT TYPE < H=HEAVY, M=MEDIUM, L=LIGHT > >

NUMBER OF ROOM TYPES IN THE BUILDING >

DESIGN OUTSIDE TEMPERATURE ON THE FIRST DAY >

INFILTRATION ALLOWANCE >

START DAY   FINISH DAY   FIRST SUNDAY

NUMBER OF DAY TYPES < MAX. 7 > >

DAY TYPE NAME
1
2
3
4
5
6
7

**END

In its simplest form it can be a calculation sheet of the type shown in Figure 1, which lists all the items required to start the calculation procedure and reduces the amount of writing to be carried out. Its only aid to numeracy is that it lays out the equations in a way that is suited to a slide rule or electronic calculator. At the next level complex equations can be solved graphically as shown in Figure 2 or tabular form as with the equivalent temperature differences shown in Table 2.

Possibly the most highly developed design aid is the computer program. Computer programs have been developed within Ove Arup Partnership to evaluate most of these parameters. STEMP, which is currently available on the HP9820 and will soon be reissued on the DEC10, calculates the environmental temperature of rooms without air conditioning so that a decision can be made about the level of servicing required.

A program is available through Arups' Mechanical and Electrical Development Group (MED) which will assess infiltration. This program is currently undergoing further development. The program in its present form takes into account wind speed and temperature difference.

Program ILLDH calculates the illumination levels on a horizontal plane due to daylighting. There are several programs to calculate illumination levels with various types of lighting fitting.

REVERB has been developed to calculate reverberation times for rooms and gives a choice of calculation method using either the Sabina or Eyring formula.

Having determined the types of systems and the internal conditions required, one must consider the capacity of these systems, which is affected both by the occupancy and the external weather conditions. Various programs have been produced to assess the external weather data. The AGNES program will plot sunpath diagrams for any location on the earth's surface. COOL calculates cooling loads for buildings whereas HEAT can be used when only heat losses are required.

One new field of study which has attracted world-wide interest is that of energy analysis. The energy program which is now available resulted from a collaborative exercise between the London and South African Arup Partnerships. This program is undergoing comparison with others and the results so far are encouraging. One other field in which simulation has proved invaluable is that of solar energy. The SOCOL program, which simulates a solar collector installation, has now been in use under supervision in its latest form for some time and has been passed to the Computer Group for inclusion in their system.

CONCLUSION

Most of the more obvious cost-effective areas have now been dealt with and future work will probably be carried out as a result of specific requests by users. If, therefore, there are any specific requirements for computer programs these can be initiated with Arups' Computer Group who will involve MED as necessary. If, however, a problem put to MED results in a computer program which seems to have general applications, then the results will be fed into the system via the Computer Group.

REFERENCES

1   Institution of Heating and Ventilating Engineers, "IHVE Guide Book A", The Institution, 1970.
2   Holmes, J M, "Calculation of Natural Ventilation Rates", MED Notes: 24, Ove Arup Partnership, 1978.
3   Ove Arup Partnership, "Oasys. Building Services Computer Programs", The Partnership, 1978.

Chapter 9

A survey of development in building design for energy economy in France

T Barker

INTRODUCTION

The energy crisis of 1974 signalled in France the beginning of what was described later as an avalanche of laws and regulations in an attempt to reduce national energy consumption. To date this avalanche has resulted in over 100 different areas of legislation covering all aspects of the French industrial and domestic scene. It is therefore surprising, and disappointing, that perhaps with the exception of new laws on thermal insulation not one of these new laws appears to have seriously affected the design concepts of buildings in France.

THE AEE

The principal source of this legislative activity is the Agency for Economies in Energy (the AEE), set up by the Government as a separately funded semi-private organisation, with the singular aim of achieving, by 1985, a national reduction of 45 million tonnes of equivalent oil consumption in energy usage.

The work of the Agency is split into three main divisions, namely:

a) industry and agriculture,
b) transport
c) residential and commercial buildings.

In the first of these sectors, the AEE has exercised its control principally by taxation legislation on the cost of fuel, the control of loans and grants, and the introduction of specific regulations, principally in industry, covering waste-heat recovery systems and heat pump development. It is now obligatory to incorporate some type of heat reclamation on factory ventilation systems exceeding two air changes an hour fresh air. In addition, minimum thermal insulation standards for factory construction have been fixed, but these stop short of fixing energy target figures, per square metre or cubic metre, as have been imposed on the domestic sector of construction.

In the transport sector, the AEE are responsible, indirectly, for vehicle speed limitations, petrol pricing, and the instigation of development work on the internal combustion engine.

It is, however, in the final category that there has been the most noticeable concentration of effort, and where the AEE has achieved their greatest success. New requirements governing lighting levels and energy consumption of domestic equipment have been introduced. Space temperatures during the heating season have been restricted to 20°C and individual thermostatic regulation is now mandatory on a per dwelling basis. In multi-dwelling buildings, individual energy metering of each unit is now compulsory, and presumably the occupier is accountable for excess energy usage above the average. Finally, up to 1977 300 million French francs had been invested on thermal insulation and control systems in existing buildings.

It is perhaps in its development section of the AEE that one can see the direction of future trends in the French building industry. Up to the end of 1976 over 76 different prototype operations had been commissioned, 30% of which were in the building sector. Among these the most significant have been:

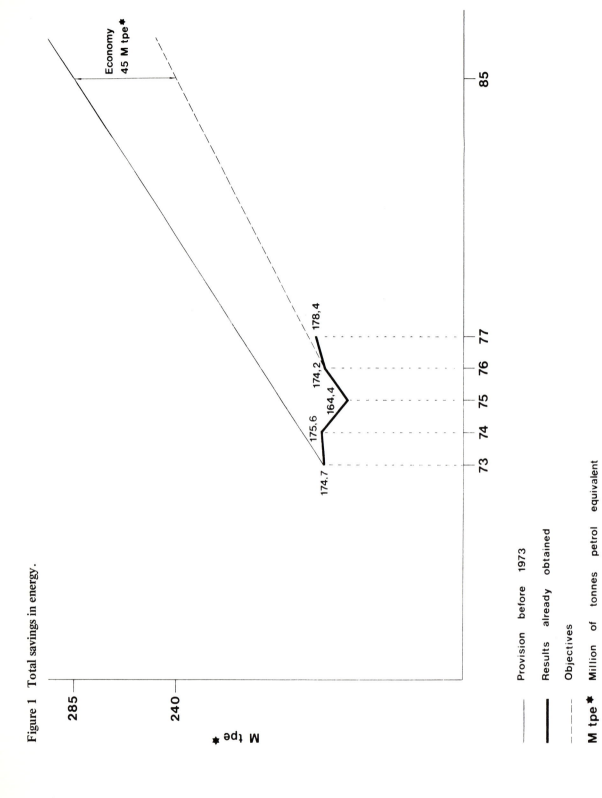

Figure 1 Total savings in energy.

Provision   before   1973

Results   already   obtained

Objectives

M tpe ✱   Million   of   tonnes   petrol   equivalent

1. The installation and monitoring of reclaim systems on industrial gases and ventilation exhaust systems.

2. The development of heat pumps to 'upgrade' and re-use waste heat.

3. Four experimental installations to examine the potential of solar collectors as a source of high temperature energy.

4. Numerous installations of steam and high temperature hot water, generated from central incineration of domestic and commercial waste products.

ECONOMIES ACHIEVED

To date it is calculated that this development work has resulted in direct energy economies of 13 400 tonnes equivalent of oil per annum, with a future potential of nearly 6 million tonnes per annum. The total savings achieved in energy terms are indicated in Figure 1. From the above it would appear that the Agency is on target to achieve its aims by 1985 but to what extent this results from the economy measures taken is debatable.

OTHER INFLUENCES

Even before the events of 1974 France had recognised the problems, both financial and political, of relying wholly on imported energy, and had embarked on an ambitious construction programme of nuclear fuelled electricity generating stations. Coupled with this the mode of heating moved away from the oil/gas fired system to the all-electric building. This was particularly evident in the public sector housing programme, where the underfloor electric background heating and electric fan convector topping up unit became the standard solution. Another 'classic' example of this policy of all-electric was the Centre Pompidou.

Pressure by the environmental protection groups resulted in a slowing down of the nuclear programme, but, judging from recent press releases, the events in Iran have again accelerated the nuclear programme and presumably, therefore, the all-electric building solution.

Another possible influence on French buildings in the future, more related to where they are built rather than what is built, is the potential of geothermal energy. Two housing projects are currently in the design phase, one of 4000 units and a second of 2200 units, where the primary heating source will be geothermal water extracted from borings, 1500 m deep at a temperature of 57°C. It is anticipated that this energy source will provide 74% of the heating and hot water energy needs of the projects, the remaining 26% being furnished by back up gas-fired systems, and the energy economy will be of the order of 4600 tonnes equivalent of oil per annum.

CONCLUSION

It is unfortunate that one cannot announce any major breakthrough either in architecture or in building technology resulting from the activity in France during the last four years. One can, however, conclude that the French engineer and his client are far more energy conscious than in yester-year and that, with a few exceptions, his design concepts illustrate this consciousness.

Chapter 10

New Italian building regulations on energy consumption

T Barker

Unlike their French counterparts (see Chapter D1), the Italian Government took until 6 February 1978 to prepare and publish their revised national regulations, governing the use of energy in buildings. These regulations are therefore some of the most recent to come out in Europe and give, perhaps, an indication of trends in Europe. The following text is intended to give a general indication of the scope of these regulations, for comparison with those of other countries.

Article 1 gives the definition of specific terms used in the regulations.

Articles 2 and 3 give the scope of application of the regulations relating to the form of heating applied in the building and the classification of the types of building as follows:

E1 domestic type residences (principal and secondary), prisons, military establishments, convents, hotels and boarding houses;
E2 office buildings;
E3 hospitals, clinics, etc;
E4 recreational buildings – cinemas, theatres, museums, libraries, churches, bars, restaurants, conference centres;
E5 commercial type buildings – shops, supermarkets, etc;
E6 sports facilities – swimming pools, etc.

Articles 4–6 set out the documentation required to obtain approval of the design proposals, including names of the approved laboratories agreed for conducting sample testing. In particular the following must be submitted for approval:

a) the heat producing plant and equipment, burners, steam generators, hot water boilers;

b) details of all equipment which is connected into the above – air/water heat exchangers, calorifiers, circulation pumps, proposed heat transfer systems, eg, radiators;

c) the type and components of the automatic control systems, and heat metering equipment.

Articles 7–16 define external design temperatures, installation requirements and maintenance instructions.

The external winter design temperature is specified for the major Italian towns, for other areas not specifically named a system of temperature correction is given.

The internal temperature is limited to 20°C ± 1°C tolerance during the hours of operation during the heating season. All installations having an installed load greater than 348 kW or 300 000 kcal/h must have not less than two boilers.

On existing installations, using liquid or gaseous fuel suppliers, systems of automatic regulation or control shall be incorporated according to the following programme dates:

Systems over 350 000 kcal/hr   : before 30 September 1978.
Systems of 250 000–350 000 kcal/h   : before 30 September 1979.
Systems of 150 000–250 000 kcal/h   : before 30 September 1980.
Systems of 100 000–150 000 kcal/h   : before 22 June 1981.

The temperature of domestic hot water is limited to 48°C, with a tolerance of 5°C.

Heating plants which serve both the heating system and the domestic hot water production system will be sized only to satisfy the building heat losses, and not take into account heat load requirements of the domestic hot water system. On centralised domestic hot water production systems each user must be separately metered.

In *Article 12* a minimum specification for the thermal insulation of pipework based on tube diameter thermal characteristics of the insulants and location in the building is given.

Articles 15 and 16 list the commissioning requirements and maintenance instructions required by the State.

Articles 17–21 are concerned with the thermal insulation of buildings and are based on two coefficients:

Cd represents the thermal transmission through the space enclosure per unit volume of the space per $^\circ$C $-$ W/m^3 $^\circ$C;

Cv represents the fresh air heating energy requirements per unit volume of air per $^\circ$C $-$ W/m^3 $^\circ$C.

Each value calculated for Cd and Cv shall be less than limit values given in the regulation. Cv may, however, take into consideration heat recuperation systems or equipment proposed which for certain types of buildings is obligatory. The coefficient Cg = Cd + Cv, and is defined as the overall thermal insulation coefficient for the building.

Articles 22–25 cover the procedure for buildings which have received planning permission prior to the issue of the new regulations, and which must be modified to conform.

ANNEXES

The following information is given in the annexes to the regulations.

1. The subdivision of Italy into six climatic zones for degree day calculations.

2. Values for Cd and Cv on a regional/climatic zone basis and also as a function of S/V where:
   V = volume of the heated spaces of the building,
   S = total surface area which encloses V.

3. Maximum air quantities and operating hours permissible in calculating Cv and requirements for heat recuperation from the ventilation/fresh air systems.

4. Degree day tables applicable to Italian towns.

5. External design temperatures.

6 and 7. Standard specifications for the operation and regulation of heating plants and tables of typical results.

Chapter 11

A Nigerian viewpoint

S Thomas

INTRODUCTION

Energy, in all its forms, is vital to sustain life. It is also fundamentally important in man's attempt to control nature.

In residential buildings, energy is used for cooking, heating, cooling, illumination and to run various household gadgets. In commercial and industrial buildings, in addition to the above, energy is essential in the manufacture and distribution of essential goods and services. The high energy demands of the world today have made us increasingly aware of the need to preserve and conserve energy as well as looking for alternative sources of energy. A deliberate policy of conservation must therefore be adopted.

To appreciate the need for energy economy in building, it is worth while to identify the various sources of energy in use today and their limitations for use in their various forms. The various sources of energy in Nigeria today are threefold:

a) traditional,
b) conventional,
c) alternative.

a) The only significant traditional source of energy is wood, which to a large extent is used in the rural areas of Nigeria for cooking. It is also used in the cities for cooking to some extent. It is cheap and satisfies the energy requirements of the rural areas of Nigeria today.

b) Conventional energy sources can be categorised as three, that is, coal, hydro-electricity, oil and gas. Coal is an important conventional source of energy in Nigeria; it has been used quite efficiently before now in the thermal electricity generating plants, as well as for cooking. But with the advent of the hydroelectricity station at Kainji this was relegated to the background either deliberately or because of the depletion of this type of fossil fuel. Natural gas as a form of energy is recent in Nigeria and was introduced to supplement the energy generated by the hydroelectric plant at Kainji and to cater for the rapidly increasing demand due to rapid industrialisation.

c) Alternative sources of energy, ie, solar energy, wind, nuclear energy and tidal or ocean waves have not been adequately explored in Nigeria, basically because the level of technology necessary to utilise these sources of energy is not available or because they are capital intensive.

GENERATION SOURCES AND GOVERNMENT POLICY

In this chapter the discussion on energy conservation in buildings will be limited to that in Nigeria, since the international concept at this point in time in Nigeria is a novelty. A brief outline of the generation of energy in Nigeria and the demand patterns will be given and put into context.

Nigeria's main source of energy, which is rather central, is Kainji hydroelectric power station. This was intended to provide the energy needs of the whole of Nigeria but with the discovery of oil the economy improved and there was rapid build up of industries requiring a sizeable amount of energy to maintain the required production level. Also, the Government is bent on keeping up the level of economic growth and the continued industrialisation of the Nigerian economy and so has embarked on a programme to generate more energy. This was done by making use of the gas, which was up to this time being fired, to generate additional energy to supplement the existing supply.

Nigeria can be termed a net exporter of energy in the sense that her vast foreign exchange earnings relative to other African states is derived from the export of fossil fuel, petroleum.

This realisation has made it difficult for Nigeria to embark on a policy of energy conservation. It should be realised, however, that Nigeria is at the moment generating far less energy than is in demand. The policy at the moment is to generate more energy by exploring other possible sources. It is, however, the sensible thing to do since 80% of this vast nation is rural and the future development of her rural population will be dependent on adequate energy generation.

ENERGY UTILISATION AND CONSERVATION IN BUILDINGS IN NIGERIA

If we look at the history of buildings in Nigeria, it can be seen that the tradition encourages energy conservation, primarily because there is a dearth of energy and the level of technology and shortage of manpower limits the growth.

Building patterns and planning in Nigeria complement conservation. This is likely to be deliberate because of the limited energy resources. In the rural areas building shapes, forms and function are still very traditional. Buildings are still being constructed in predominantly clay mud, which cools the building naturally in the hot season and warms it in the cold season. The disadvantage, however, in the use of this type of material is the poor natural lighting. A lot of the domestic energy requirement is, however, still being provided for by the burning of coal or wood.

At the end of the colonial era traditional buildings were giving way, especially in the city centres, to European building patterns. The forms and patterns, even at that time, also kept the energy requirements of building to the bare minimum or at least allowed for alternatives. For example, buildings in the senior service quarters in Lagos were built with two kitchens, one deriving its energy requirements from wood, the other from the limited electrical source generated from coal. The rooms were properly ventilated by providing large areas for cross-ventilation with movable wooden louvres the entire height of the storey. This also served the other role of allowing a very high proportion of the natural lighting into the rooms and in fact planning permission was based on these requirements being fulfilled.

The discovery of oil and the prosperity that comes with it saw a large influx of foreigners into Nigeria, bringing with them ideas foreign to her culture and need. With Nigeria's quest for modernisation a lot of buildings were constructed with very high energy requirements, although the architects were quite aware of the inadequacies of the country's energy resources; it was then a question of "I have got something to sell and you can afford it, it does not matter if you do not need it now". Because the immediate luxury and beauty of this foreign architecture was desired by Nigerians and the country wanted to be international, a lot of the original standards and requirements were overlooked and a great deal of energy was required in buildings to provide cooling, lighting and other gadget requirements.

Nigeria is now faced with the realisation that her newly discovered energy source, which was primarily sold to earn the much needed foreign exchange, is after all depletable and coupled with her limited technological base in utilising the gas from her oil to generate more energy. A deliberate return to the simpler form of life brought about by minimising the energy requirements in buildings must be sought but is seems unlikely that this will happen until there is a government policy on it in Nigeria.

Consultants in Nigeria are now embarking, unintentionally though, on energy conservation because there is a tendency for clients to put a ceiling on the project cost. The mechanical engineers, for instance, are now demanding architectural elevations that will reduce the heat gain in buildings so that a simple form of cooling system can be utilised.

Sometimes they have also specified that in office buildings there should be separate air-conditioning units on each floor. The efficacy of this cannot be defended, but they claim that the energy requirements can be controlled per floor or the air conditioning at least switched off on floors when it is not required at any particular time. They have also gone to the extent, in some cases, of using simple equipment such as the extractor fan, which supplies the fresh air intake into a building at a low level, and the exhaust fan, which expels hot air at a high level without the need for cooling. This, they know, has its limitations but at least supplies the need within the budget.

MODERN TRENDS OF CONSERVATION IN BUILDINGS INTERNATIONALLY (with particular reference to the developed nations)

Energy demands of the world today are ever increasing, both in the industrialised and developing countries. At the time when only a few countries were developed, developed countries had free access to the world's energy reserves, and there was no shortage of energy. In recent times a lot more countries have become independent and consequently there has been the drive for development which has increased the world energy consumption.

In view of the sharp increase of oil prices, the world was shocked into the realisation that the stock of fossil fuels is in fact exhaustible. There was therefore a deliberate policy by the industrial nations to conserve energy. It is interesting to note and recall what has been achieved in these countries in relation to buildings.

Buildings shapes, forms, construction and orientation have been influenced to an extent. There has been a considerable study of the thermal capacities of buildings; designers have addressed their minds to shading buildings, providing natural ventilation and lighting and on the whole reducing the energy requirements. They have also resorted to minimising wastage as far as practicable by studying the transfer of energy in all its forms and assessing the percentage efficiency of machines. More than ever before they have intensified their search for new energy sources.

CONCLUSIONS

It is the author's opinion that history will repeat itself, in the sense that technological awareness took a certain course in its utilisation and production of energy, and so will conservation. It is to be hoped that the world will be willing to return to the simpler form of life in the desire to conserve energy, or at least to intensify efforts in the search for utilisation of solar energy which is inexhaustible and free.

Chapter 12

Total systems design in air conditioning and lighting

H Spoormaker

INTRODUCTION

When we consider future trends in building design and the need for energy conservation, we must always remember that the highest cost is the salaries of, and equipment for, people working in the buildings. It does not make much difference how you look at it — accumulated total cost over 40 years of the building life, or present day value with discounted future annual cost. In Table 1 accumulated cost is shown on the left and present day value on the right. It is therefore essential to maintain the utilisation efficiency of the building, as the initial cost of the building is only a very small percentage of the total cost.

Depending on the design philosophy there are roughly three approaches for designing buildings. Firstly, and, at present, becoming quite popular, is 'minimum is best'. The last one — shown in the right hand column of Table 2, 'more is better' — is rapidly becoming outdated. We believe that the best solution is between these two extremes, indicated as 'total systems approach'. It is the optimum balance between a central 'de luxe' air conditioning system and ignoring the comfort control altogether. It would consist of a central plant system for those functions which should be done centrally (ie humidity, dust, fresh air control), and factory-produced self-contained air conditioning units, and room fan air terminals, with a floor plenum for distribution and flexibility.

ESCALATION OF ENERGY COST

The expected escalation of energy cost will play an important role in future design of buildings. Our present experience which is used in designing buildings is based on the period 1950 to 1975 which is indicated as 'experience' on Figure 1. During this period the relative cost of electricity doubled. We have also indicated as 'expectation', how the energy cost will rise during the next thirty years. There is general agreement, at least in Europe and the USA, that energy cost will rise at a rate equal to general escalation plus 5%. As can be seen, our experience of the past does not have much relevance for the future.

The present range of electricity consumption in kilowatt hours per square metre per annum for various types of office buildings is considerable. Before the energy crisis, energy conservation was no criterion for design and it could therefore vary from 180 to 600 kilowatt hours per square metre per annum; the majority of the 'prestige type' buildings are in the order of 400 to 500. In 1975 various codes for energy conservation in Europe and the United States were published and the prescribed energy consumption levels are indicated by the lower, almost horizontal line in Figure 2. With a total systems approach, energy consumption levels below these codes for energy conservation can be achieved as indicated by the lowest line.

DESIGN FOR COMFORT CONTROL

Figure 3 indicates schematically how optimum comfort control can be achieved in two ways. The first is by using a properly designed building envelope to eliminate high heat gains and losses, hot and cold radiation, rapid temperature changes, draughts and infiltration. In a number of modern buildings the air conditioning system has to perform a corrective function because, for various reasons, the architect has ignored the functions of

Table 1 Total cost of an office building and its occupants.

	ACCUMULATED COSTS OVER 40 YEARS OF BUILDING LIFE	PRESENT DAY VALUE FUTURE & INITIAL COST (8 - 10% INTEREST) (10 - 15% ESCALATION)
SALARIES AND OFFICE EQUIPMENT	80 - 90%	70 - 80%
ENERGY OPERATORS MAINTENANCE	15 - 7%	15 - 10%
BUILDING AND SYSTEMS	5 - 3%	15 - 10%

Table 2 Effects of design philosophy on building design.

FUTURE TRENDS

DESIGN PHILOSOPHY	MINIMUM IS BEST	TOTAL SYSTEMS APPROACH	MORE IS BETTER
BUILDING ENVELOPE	LOWEST CAPITAL COSTS	ENERGY CONSERVATION	AESTHETIC AND IMAGE
COMFORT CONTROL SYSTEM	NO { DUST NOISE DRAFT } CONTROL MINIMUM HEATING THERMAL STRESS RELIEVED THROUGH OPEN WINDOWS OR WINDOW UNITS	DUST - NOISE - INFILTRATION CONTROL BY PRIMARY AIR CENTRAL BUILDING SYSTEM CENTRALLY PROGRAMMED HEATING SYSTEM FINAL COMFORT CONTROL BY PERSONALLY CONTROLLED FLOOR AIR SUPPLY AND BOOSTER COOLERS	"DE LUXE" AIR CONDITIONING WITH MODULAR (AND CONCEALED) AIR DIFFUSERS AND (HIDDEN) THERMOSTATS
ILLUMINATION	MINIMUM	ENGINEERED AS FUNCTION OF WORK STATION ACTIVITY	MODULAR GENERALLY HIGH LEVEL FOR FLEXIBILITY
TOTAL SYSTEMS PRESENT DAY COST	HIGH	LOWEST POSSIBLE	HIGH

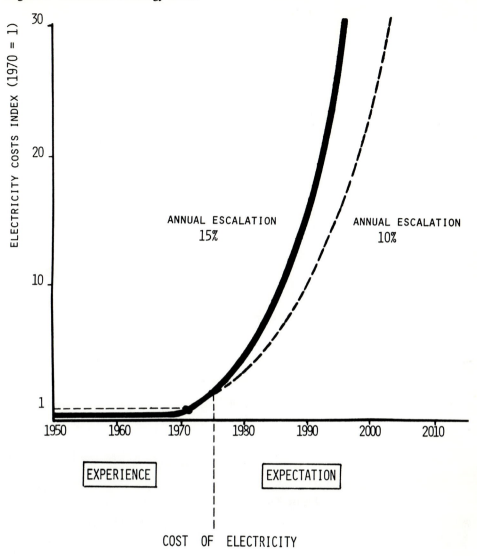

Figure 1 Escalation of energy costs.

the building envelope. Secondly, control can be achieved by using the air conditioning system to deal with the function of final comfort trimming, which cannot be provided by even the best designed building envelope. These functions are as follows: final temperature selection, room flushing, humidity control, dust and noise control.

LIGHTING

The electric power required for the illumination systems is a major factor in total building energy cost. Figure 4 indicates how electric lighting loads first increased from 1950 to the mid-seventies and are now moving down to the energy levels of the 1950s. During the period of increase of load, however, a basic improvement of the quality of the light fittings took place and the present low energy levels can be achieved with much better illumination than that of the original ceiling, surface mounted bare fluorescent tubes.

Because energy reduction is linked to task illumination it is essential to examine the actual task illumination requirements in designing the illumination system. Figure 5 indicates how to select the illumination level depending on the type of work or activity in the space (indicated on the left-hand side) in order of increasing requirements:

a) circulation area — stairs, entrance hall;
b) casual short work — canteens;
c) routine work — executive offices, conference rooms, classrooms;
d) average routine work — offices, kitchens, shops;
e) demanding work — drawing or business machine offices.

Figure 2   Energy consumption in office buildings.

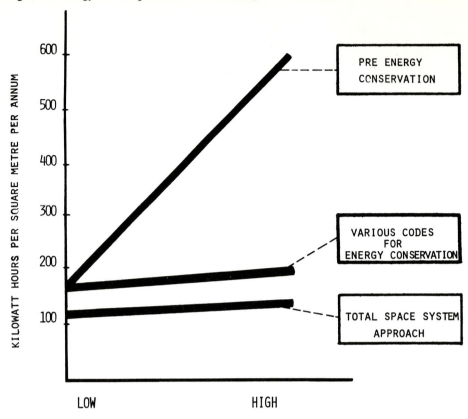

Figure 3   Comfort control.

OPTIMUM COMFORT CONTROL

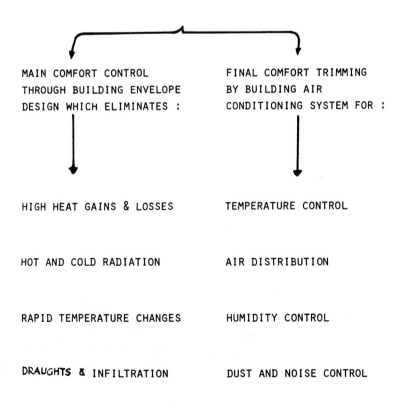

Figure 4 Development of illumination systems, 1950–1980.

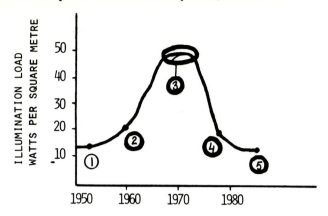

1. CEILING SURFACE MOUNTED BARE FLUORESCENT TUBE

②. CEILING SURFACE MOUNTED TASK RELATED GOOD QUALITY LIGHT FITTINGS

③. HIGH ILLUMINATION LEVEL MODULAR CEILING RECESSED GOOD QUALITY LIGHT FITTINGS

④. RECESSED CEILING MOUNTED TASK RELATED HIGH EFFICIENCY LIGHT FITTINGS

⑤. AMBIENT AND TASK LIGHTING ENGINEERED AS A FUNCTION OF THE WORK STATION ACTIVITIES

Figure 5 Illumination flow chart.

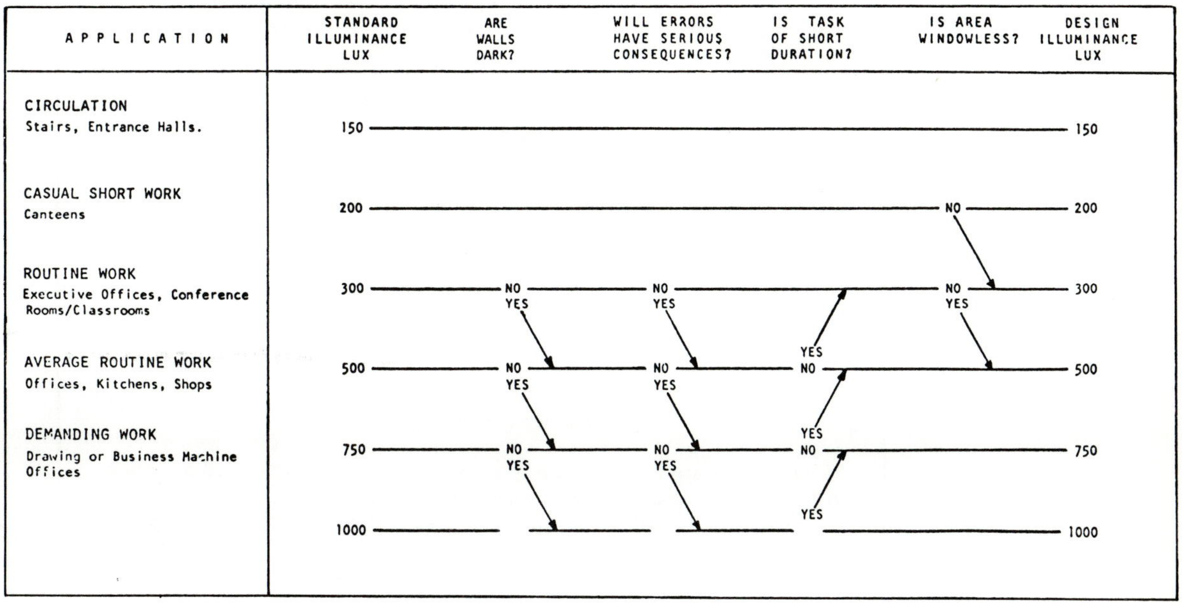

APPLICATION	STANDARD ILLUMINANCE LUX	ARE WALLS DARK?	WILL ERRORS HAVE SERIOUS CONSEQUENCES?	IS TASK OF SHORT DURATION?	IS AREA WINDOWLESS?	DESIGN ILLUMINANCE LUX
CIRCULATION Stairs, Entrance Halls.	150					150
CASUAL SHORT WORK Canteens	200				NO	200
ROUTINE WORK Executive Offices, Conference Rooms/Classrooms	300	NO / YES	NO / YES	YES	NO / YES	300
AVERAGE ROUTINE WORK Offices, Kitchens, Shops	500	NO / YES	NO / YES	YES	NO	500
DEMANDING WORK Drawing or Business Machine Offices	750	NO / YES	NO / YES	YES	NO	750
	1000			YES		1000

This is followed by a column listing the average standard illumination levels. These have to be corrected depending on the following criteria:

1. Are walls dark?
2. Will errors have serious consequences?
3. Is the task of short duration only?
4. Is natural lighting available?

The arrows indicate how, in many cases, the illumination has to be increased and, in certain cases, can be decreased depending on the answer to these questions. From this table the correct design level of illumination can then be selected.

The quality of lighting — glare control and reflections — is important, especially with task lights. Figure 6, which is an extract from the General Services Administration in the USA, indicates how the light fitting must be positioned in the ceiling relative to the work station if the task illumination is provided by ceiling-mounted light fittings.

Another solution is a combination task and ambient lighting scheme which uses free floor standing lights and light fittings in or on the furniture. Figure 7 shows task and ambient lighting integrated in the furniture for a secretarial work station in the Arco Building in Philadelphia, USA. There is no lighting in the ceiling.

Figure 6 Positioning of ceiling-mounted light fittings.

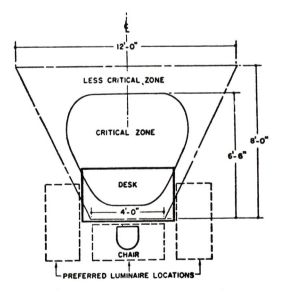

Guide for Non-Uniform Office Lighting Layouts
with Direct Luminaires

1.   Locate work surfaces and determine lighting needs for tasks. Desks, tables, credenzas, files, etc., will have tasks. Determine their difficulty, specularity, and the plane on which they lie.

2.   Within the limits of the luminaire supporting system, locate the luminaires as close to directly over the task as possible without creating excessive veiling reflections. With highly specular tasks -- shiny paper, pencil writing -- try to avoid the trapezoidal ceiling area defined by (for a 9'-0" ceiling height): the working edge of the desk projected vertically to the ceiling forming a line 4' long centered on the worker, a line parallel to and 8' forward of this line that is 12' long, the sides of the trapezoid connecting the ends of these two base lines. See drawing below.

12'-0"
LESS CRITICAL ZONE
CRITICAL ZONE
8'-0"
6'-6"
DESK
4'-0"
CHAIR
PREFERRED LUMINAIRE LOCATIONS

If luminaires are kept out of this zone, the LEF* should be 1.0 or better. If part of a luminaire projects onto the edge of this area, only minor visibility losses will occur, producing an LEF on the order of .8 to .9. If the bulk of a luminaire or parts of several luminaires project onto this zone, particularly near the desk, LEF can be .4 to .7. Higher

Figure 7   Integration of
lighting with furniture.

Figure 8   Free-standing light fittings.

Figure 8 is an example of a task light fitting as well as an ambient light fitting, which is free standing and connected to the distribution system in the floor.

Figure 9 shows an open plan office suite with a task and ambient lighting illumination system consisting of:

a) vertical, floor standing, indirect ambient light fittings;
b) task and ambient light fittings mounted on the space dividers and acoustic screens;
c) fully adjustable task light mounted on the desk. The task light, which can be swivelled in any position to control the glare, can be used on any standard type of desk.

Figure 9   A combination of lighting types.

AIR CONDITIONING

After the illumination design criteria have been considered the factors which influence the air conditioning design criteria have to be examined.

Figure 10 shows the statistical different behaviour between males and females. It illustrates how the different sexes adjust the insulation value of their clothing depending on the average external temperature during the week — left is winter, right is summer. The graphs show that both males and females are in agreement as far as their clothing is concerned when it is very cold outside. However, they differ substantially when it becomes warmer. The graph shows that females have a more sensible clothing adjustment policy than males. It is clear from this graph why it is never possible with conventional average air conditioning to satisfy both males and females in the same space. If you try to satisfy both, the females will complain of being too cold and the males will complain of being too hot.

The ideal comfort temperature experienced by people is dependent on their clothing insulation, their activity, the space temperature and local air movement.

Figure 10 Difference in clothing variation between males and females.

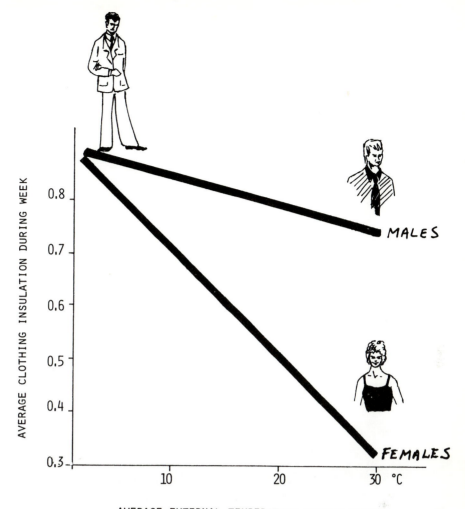

Table 3 shows the variation in temperature for various activities depending on the clothing level. In the lower part of the table the possible increase in space temperature is shown as a function of the increase in local air movement. The traditional 'de luxe' air conditioning design condition for offices is 23°C for 'sitting' activity, and 'normal' clothing.

This temperature can be increased to 23°C + 2.5 = 25.5°C if local air movement is increased to 1 m/s. Further adjustment of a few degrees is possible if clothing is adjusted from 'normal' to 'cool'. To conserve energy in the USA systems are now set at 26.5°C in the summer and 20°C in winter.

Description of various types of clothing: 'nude', 'light' and 'normal', is shown in the lower section of Table 3.

Figure 11 shows the principle and some examples of structural storage cooling. On the top graph we have shown how the outside air temperature varies much more than the inside temperature and, if the inside temperature variation stays within the comfort level, then theoretically no cooling is required.

The first example (1) of structural cooling is a site office in a building under construction. The air flow over the person was regulated by varying the opening of a cover which covered a hole between the office and a cool basement. The basement was cooled at night by the outside air drawn in at ground level to replace the hot air in the building which was expelled at the top through stratification. The occupant controlled his comfort by varying the cover and using an exhaust fan. The position of the desk was relatively close to the opening so as to use local air movement from the opening over the head of the occupant.

Table 3 Comfortable room temperatures (°C) in still air with corrections for air movement. *(Taken from Building Research Establishment Digest June 1979, Thermal, Visual and Acoustic Requirements in Buildings, and reproduced by courtesy of BRE)*

CLOTHING	ACTIVITY			
	SLEEPING	SITTING	STANDING	ACTIVE
NUDE	31	29	25	28x
LIGHT	29	26	· 21	18x
NORMAL	27	23	17	13x xx

AIR MOVEMENT m/s	CORRECTION TO BE ADDED			
0.2	0.5	1	1	1.5
0.4	1	1.5	2	3
0.7	1.5	2	3	4
1.0	1.5	2.5	3.5	5

x   Estimates subject to some uncertainty

xx   Below the legal requirement for offices and factories.

CLOTHING CATEGORIES

NUDE	Naked, or with light underwear, bikini or bathing trunks. Clothing insulation : 0.0 - 0.2 clo
LIGHT	Light summer dress, skirt and blouse, shorts and shirt, trousers and shirt. (0.3 - 0.7 clo)
NORMAL	Winter dress, skirt and jumper, trousers and jumper, three piece suit. (0.8 - 1.2 clo)

Another example (2) is an office in West Germany which was conditioned without any refrigeration. The main building extract fan drew the cool night air through the space between the strucutral slab and the ceiling, using flaps in the facade, which were opened during the night. In the day time flaps were closed and the main extract fan was switched off. If cooling was required, cooling fans in the ceiling were switched on to draw the warm air from the room, discharge it against the cool structure and then return it back to the offices through slots in the ceiling.

Pre-cooling the building structure at night can reduce the installed capacity and running cost. Figure 12 shows the instantaneous power demand during various times of the day for three different buildings and air-conditioning systems. Energy demand or maximum demand is an important factor as it determines 50 to 60% of the total cost of electricity.

The top curve is for a prestige type energy inefficient building and system. Energy conservation was not a consideration in the design. This could be improved as indicated by the second lower curve through a more energy efficient building envelope and energy efficient system, operated with energy conservation in mind. However, the major saving is achieved when a structural storage system with night cooling is put into an energy efficient building — indicated by the lowest curve.

All these factors highlight the need for re-examining the present normal air conditioning design standards. In Table 4 we have indicated three possible standards, the 'de luxe', shown in the left-hand column, being the conventional way of designing air-conditioning systems. It is suggested that serious consideration should be given to 'normal', which can provide a better comfort experience for the occupants, with higher temperatures, provided that people are allowed to adjust their clothing and can adjust local air move-

Figure 11   Structural storage cooling.

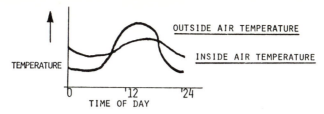

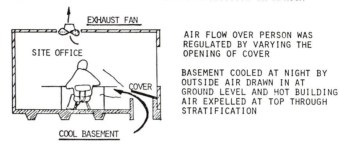

1.   OFFICE IN BUILDING UNDER CONSTRUCTION IN AFRICA

EXHAUST FAN

SITE OFFICE

COVER

COOL BASEMENT

AIR FLOW OVER PERSON WAS
REGULATED BY VARYING THE
OPENING OF COVER

BASEMENT COOLED AT NIGHT BY
OUTSIDE AIR DRAWN IN AT
GROUND LEVEL AND HOT BUILDING
AIR EXPELLED AT TOP THROUGH
STRATIFICATION

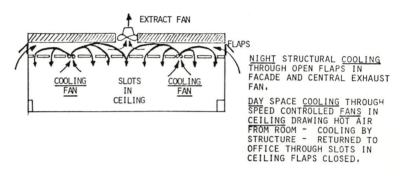

2.   OFFICE BUILDING IN WEST GERMANY

EXTRACT FAN

FLAPS

COOLING FAN   SLOTS IN CEILING   COOLING FAN

NIGHT STRUCTURAL COOLING
THROUGH OPEN FLAPS IN
FACADE AND CENTRAL EXHAUST
FAN.

DAY SPACE COOLING THROUGH
SPEED CONTROLLED FANS IN
CEILING DRAWING HOT AIR
FROM ROOM - COOLING BY
STRUCTURE - RETURNED TO
OFFICE THROUGH SLOTS IN
CEILING FLAPS CLOSED.

Figure 12   Design day demand profiles for three different buildings and air-conditioning systems.

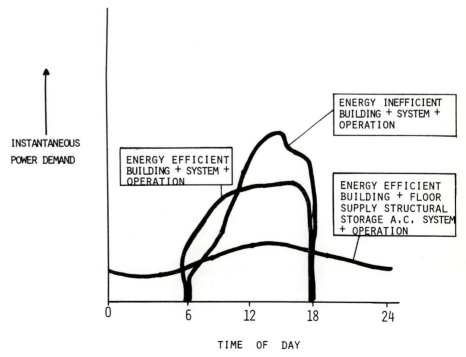

Table 4   Air conditioning design standards.

	DE LUXE	NORMAL	ECONOMY
MAX SUMMER TEMPERATURE	23°C	24.5°C	26°C
MIN WINTER TEMPERATURE	21°C	20°C	18°C
AIR TEMP VARIATION DURING DAY	2°C	3°C	4°C
NECESSARY CLOTHING VARIATION DURING DAY	NONE	WARM → AVERAGE or AVERAGE → COOL	WARM → COOL
NECESSARY AIR MOVEMENT VARIATION DURING DAY	NONE	0.1 - 0.7 M/s	0.1 - 1.0 M/s

ment. If we select 'economy' it may be possible, in areas that have hot climates but also have the benefit of cool nights, to eliminate the need for refrigeration altogether for certain buildings.

It is interesting that economy design standards are presently prescribed in the USA for office buildings, to conserve energy. However, it is thought that people will experience discomfort if economy design is adopted and that there will be a need to wear light clothing and use local air movement by means of desk fans to avoid this.

Air conditioning design is not only a question of building physics and physiological principles, but also a question of analysis of the usage of the space. Cooling loads vary from area to area on a typical floor, depending on the space function and the time of the day. This is illustrated in Figure 13. The plans on the left indicate how a boardroom located on the perimeter can have a daily room load variation of 16 (full cooling to virtually shut down).

The boardroom, which in 1979 may be positioned in the corner, could be repositioned in 1980 in the middle of the floor. The variation in maximum load between a boardroom and an average office is by a factor of about 2. The air-conditioning system must cater for these shifting loads caused by variation in occupation during the day and relocation on the floor from year to year.

To save energy and capital cost we must aim for minimum mechanical plant capacity. This is achieved if:

1. The air flow is reversed from the conventional ceiling to floor to from floor to ceiling, resulting in a possible reduction in capacity of 30%.

2. The average building temperature is allowed to rise during the day.

Figure 13 Space and air-conditioning flexibility.

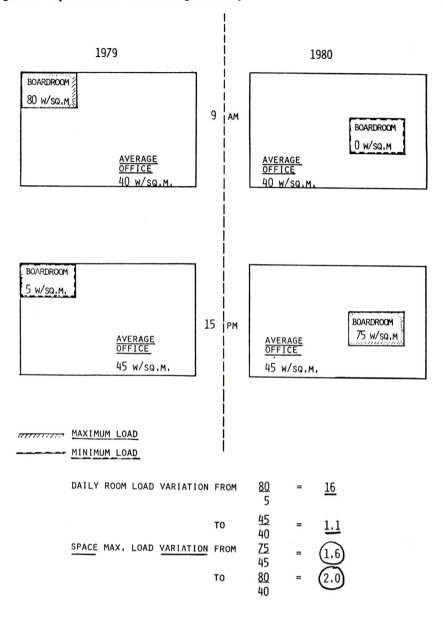

3. The general building pre-cooling/preheating is done at night.

4. Energy losses in distribution systems are reduced to the minimum.

5. Heat pump-type sensible space coolers and heaters are used to satisfy the perimeter heating and cooling requirements in excess of those provided by the relatively small central plant.

This is schematically indicated in Figure 14.

Figure 15 shows a type of building and air conditioning solution which should be able to meet the six desirable criteria of:

a) minimum mechanical plant capacity,
b) minimum energy cost,
c) personal comfort,
d) simple design and installation,
e) air conditioning flexibility,
f) easy operation and maintenance.

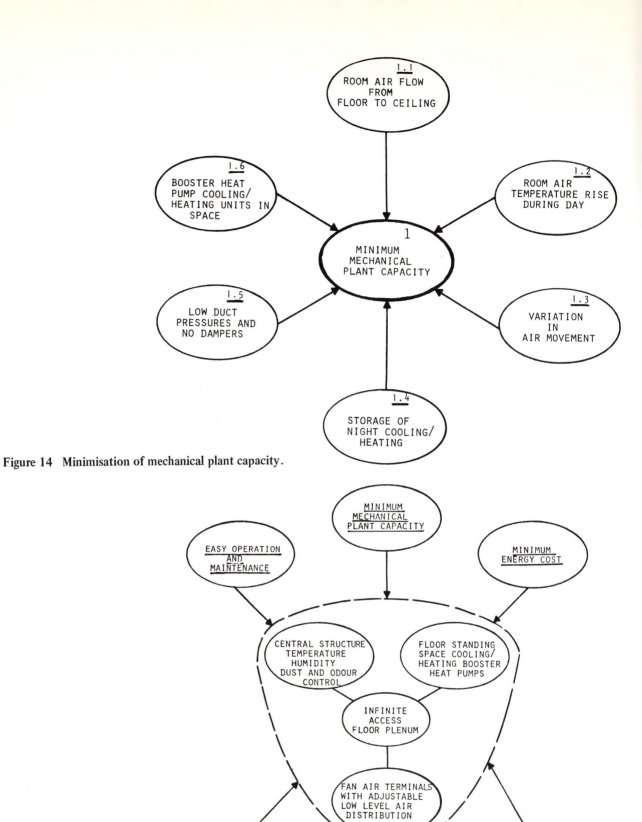

Figure 14   Minimisation of mechanical plant capacity.

- 1.1 ROOM AIR FLOW FROM FLOOR TO CEILING
- 1.6 BOOSTER HEAT PUMP COOLING/ HEATING UNITS IN SPACE
- 1.2 ROOM AIR TEMPERATURE RISE DURING DAY
- 1 MINIMUM MECHANICAL PLANT CAPACITY
- 1.5 LOW DUCT PRESSURES AND NO DAMPERS
- 1.3 VARIATION IN AIR MOVEMENT
- 1.4 STORAGE OF NIGHT COOLING/ HEATING

Figure 15   Ideal air-conditioning criteria.

- MINIMUM MECHANICAL PLANT CAPACITY
- EASY OPERATION AND MAINTENANCE
- MINIMUM ENERGY COST
- CENTRAL STRUCTURE TEMPERATURE HUMIDITY DUST AND ODOUR CONTROL
- FLOOR STANDING SPACE COOLING/ HEATING BOOSTER HEAT PUMPS
- INFINITE ACCESS FLOOR PLENUM
- FAN AIR TERMINALS WITH ADJUSTABLE LOW LEVEL AIR DISTRIBUTION
- AIR CONDITIONING FLEXIBILITY
- PERSONAL COMFORT
- SIMPLE DESIGN AND INSTALLATION

Table 5 Comparison of air-conditioning systems.

	NEW	TRADITIONAL
AIRFLOW	STRATIFICATION COOL - LOW HOT - HIGH	MIXED AIR UNIFORM TEMPERATURE IN SPACE
COMFORT CONTROL	INDIVIDUAL MANUAL CONTROL OF TEMPERATURE + AIR MOVEMENT	AVERAGE ROOM THERMOSTAT CONSTANT PRESET AIR MOVEMENT
AIR DISTRIBUTION SYSTEM	ZERO PRESSURE PLENUMS WITH FAN TERMINALS	MEDIUM PRESSURE DUCTING WITH AIR TERMINALS AND DAMPERS
CRITERIA FOR INSTALLED CAPACITY	AVERAGE DAILY LOAD	INSTANTANEOUS DEMAND
MAINTENANCE	"PLUG-IN" SMALL ALL ELECTRIC COMPONENTS. REPLACED AND REPAIRED IN WORKSHOP	CONTROL DEVICES INTEGRAL IN DISTRIBUTION SYSTEM ADJUSTED AND REPAIRED IN-SITU

This is achieved by having a system that combines central structure temperature, humidity, and dust and odour control working in conjunction with floor standing space cooling or heating booster units. These small self-contained standard units reject or extract heat in or from a condenser water piping system. They do not need fan speed control, and use a floor plenum for air distribution. The central plant and self-contained heat pumps work together through the floor plenum formed by an infinite access floor. Final air distribution is controlled by the fan air terminals with adjustable low-level air distribution.

In Table 5 we have listed how this new type of air-conditioning system differs from the conventional in the aspects of air flow, comfort control, air distribution system, criteria for installed capacity and maintenance.

SYSTEMS DESIGN

The schematic arrangement in Figure 16 shows the interaction between the floor plenum and the office space above. The work station is complete with a task light, telephone and power outlet and a desk air outlet. The air is supplied at the right humidity, cleanliness and temperature level by a central primary air supply, indicated in the rear, which can be integrated in the building structure. If the central primary air supply is not sufficient, then the space booster cooling/heating unit, indicated on the right — standing on the floor and discharging into the plenum — is installed. This unit is connected to the water pipes running in the floor plenum.

The cool air, which is cooled by the primary air interacting with the structure and, if necessary, further cooled by the space cooler, is drawn off by the fan air terminal and either discharged straight into the room or distributed by means of flexible connections to floor outlets and work station outlet as indicated. The electrical system consists of electrical conduits laid on the floor, terminating in terminal boxes. From these boxes, power supply is plugged into ambient, floor-standing light fittings, task lights on the desk, a power outlet on the desk and a space cooling unit and air-conditioning fan air terminal.

79

Figure 16 Interaction of floor plenum and office space.

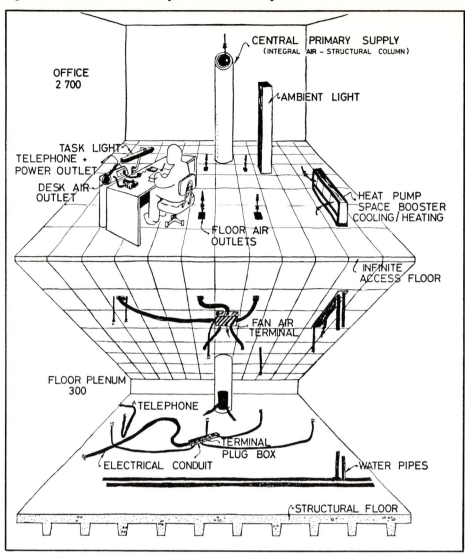

Figure 17 Flexible total-space system using an access floor.

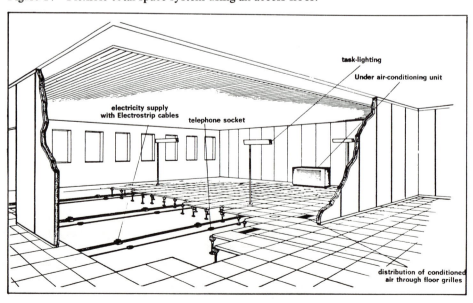

Figure 18 Detailed view of the floor plenum.

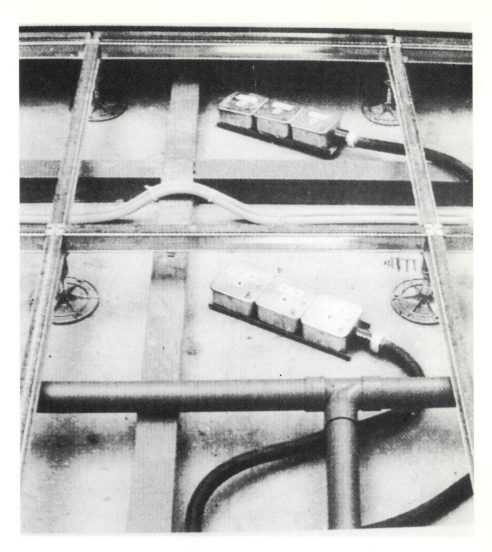

Figure 19 An intensively utilised space.

Figure 20 'Pulse' unit.

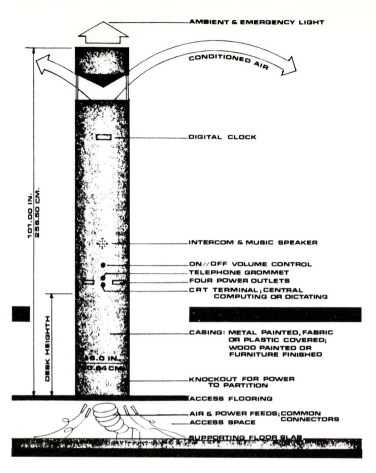

Figure 17 is another schematic drawing showing a range of possible components for a flexible total-space system using an access floor. It shows an electric and distribution system and air distribution terminals in the floor. Above floor level there are free-standing task lights and a console type air-conditioning unit recirculating and cooling the room air and discharging it into the ceiling plenum below. Figure 18 shows a view of the air plenum, electrical distribution with terminal plug box, water piping, trunking for telephone, pneumatic connection and the floor supporting grid on which the tiles are laid.

Figure 19 is an overall view of a rather intensively-utilised space, where the lighting is provided by a combination of task lights on the desk, task lights in the space dividers and free-standing ambient light fittings throwing light onto the ceiling. Air is provided through the infinite access floor.

In the USA, changes in office space systems are not as drastic as in Europe. However, changes are also taking place, as Figure 20 illustrates. This shows a pulse unit, providing lighting, conditioned air, clock, music, intercom, volume control, telephone, power outlets, electronic CRT terminals and central computing.

ECONOMIC ANALYSIS

In the past a building and its air-conditioning and illumination systems, etc were considered as the total building. However, to do a correct economic analysis and building and system comparison, it is necessary to split the building into two parts, as indicated in Table 6, ie:

   a) inside systems and,
   b) basic building.

Table 6 Necessary subdivision of total building for analysis of life cycle cost and systems selection.

MAIN SUB DIVISION	INSIDE SYSTEMS	BASIC BUILDING
ELEMENTS OR SYSTEMS	AIR CONDITIONING ILLUMINATION POWER & TELEPHONE CEILING FLOOR SPACE DIVIDERS	FACADE ROOF STRUCTURE CORE FOUNDATIONS
COST PARAMETER	PRESENT DAY VALUE EQUALS INITIAL CAPITAL COST PLUS DISCOUNTED FUTURE COST OF ENERGY – STAFF MATERIAL	INITIAL CAPITAL COST
LIFE CYCLE	1/3 – 1/2 BUILDING LIFE 15 – 25 YEARS	BUILDING LIFE 40 – 60 YEARS

The basic building, which consists of facade, roof, structure, core and foundation, has a life of 40 to 60 years. The cost factor used for that part of the building is initial capital cost only. Inside systems, which cover air conditioning, illumination, power and telephone, distribution systems, ceiling, floor and space dividers, have a life of 15 to 25 years, which is roughly a third to half the building life. The cost factor used for these inside systems is present day value which is initial capital cost plus the discounted future cost of energy, staff and material.

The following formula is used to determine the value of the discounted future annual cost with annual escalation and for a certain interest rate.

$$\text{PDV} = \sum_{1}^{N} A \frac{(1 + E)^N}{(1 + I)^N}$$

PDV	=	Present day value
A	=	Specific annual cost
N	=	Number of years
E	=	Escalation rate (decimal)
I	=	Interest rate (decimal)

The results of this formula for certain escalation and interest rates and number of years given an annual cost multiplier. Multiplying the present annual cost with this multiplier determines the present day value of this annual cost. The various multipliers, depending on the number of years, escalation rate and interest rate are shown in Table 7.

Table 7 shows that, if the number of years is taken at 40, and we deal with an escalation rate of 15% and an interest rate of 10%, we can prove almost anything, because the multiplier is as high as 113. It is considered to be more realistic to work on a life cycle of 15

Table 7 Multipliers to determine present day value of an annual cost for different escalation and interest rates and number of years.

NUMBER YEARS		15		25		40	
ESCALATION RATE		10%	15%	10%	15%	10%	15%
MULTIPLIER	8% INTEREST	17.4	25.7	32	62.5	59.6	186
	10% INTEREST	15	21.8	25	46.9	40	113

Table 8 Comparison of costs of two space systems approaches.

COST OF SYSTEMS	FLOOR/CEILING — ILLUMINATION / H.V.A.C. (MULTIPLIER)	ADAPTABLE CEILING — CEILING AMBIENT TASK LIGHTS / CEILING AIR TERMINALS		INFINITE ACCESS FLOOR & PLENUM — FLOOR AMBIENT & TASK LIGHTS / FAN AIR TERMINALS IN OR ON FLOOR	
AIR CONDITIONING:					
1. CAPITAL COST		55	55	40	40
2. ANNUAL COST OF					
– ENERGY	22	2.2	48	1.5	33
– STAFF/MAINTENANCE	15	2.5	37	2.0	30
– SPACE CHANGES	15	0.8	12	0.2	3
PRESENT DAY VALUE			152		106
ILLUMINATION & POWER TELEPHONE:					
1. CAPITAL COST		35	35	30	30
2. ANNUAL COST OF					
– ENERGY	22	2.5	55	1.5	33
– RELAMPING	15	1.0	15	0.6	9
– SPACE CHANGES	15	0.8	12	0.3	5
PRESENT DAY VALUE			117		77
CEILING:					
1. CAPITAL COST		15	15	8	8
2. ANNUAL COST OF					
– SPACE CHANGES	15	0.3	5	0	
PRESENT DAY VALUE			20		8
FLOOR:					
1. CAPITAL COST				32	32
PRESENT DAY VALUE			0		32

TOTAL CAPITAL COST R/SQ.M   105   110
TOTAL PRESENT DAY VALUE R/SQ.M   289   223

PRESENT DAY VALUE MULTIPLIER BASED ON :

15 YEARS	SYSTEMS LIFE
10%	INTEREST RATE
15%	ENERGY ESCALATION RATE
10%	STAFF AND MATERIAL ESCALATION RATE

years for the systems; this still gives a multiplier of about 22 for energy escalation at an annual interest rate of 10%. If we use a general inflation rate of 10% for normal staff, material and energy, we have a multiplier of 15. These two values are used on the next analysis.

Table 8 compares the cost of two types of space systems approaches. The one on the left is for a standard good quality commercial office building with an adaptable ceiling, ceiling ambient task light and ceiling air distribution, while the one on the right is the new type of system with an infinite access floor, floor standing ambient and task lights and fan air terminals for final air distribution.

The initial capital cost is indicated in the left column which gives a total capital cost of 105 for the conventional building and 110 for the building with the access floor. The second column gives the annual cost of energy, maintenance and modifications, and in the third column the values of these annual costs when multiplied with the present day value multiplier. The table shows that the present day value of the conventional good quality commercial inside system is 289, compared with 223 for the new type of inside system.

Table 9 Study for a building with console air conditioning units.

INSIDE SYSTEMS / COST OF SYSTEMS	FLOOR/CEILING (ILLUMINATION, H.V.A.C.) MULTIPLIER	ADAPTABLE CEILING — CEILING AMBIENT TASK LIGHTS / CONSOLE AIR CONDITIONING UNITS		INFINITE ACCESS FLOOR & PLENUM — FLOOR AMBIENT & TASK LIGHTS / FAN AIR TERMINALS IN OR ON FLOOR	
AIR CONDITIONING:					
1. CAPITAL COST		30	30	35	35
2. ANNUAL COST OF					
– ENERGY	22	4.5	99	1.5	33
– STAFF/MAINTENANCE	15	2.7	40	2.0	30
– SPACE CHANGES	15	0.2	3	0.2	3
PRESENT DAY VALUE			172		101
ILLUMINATION & POWER TELEPHONE:					
1. CAPITAL COST		35	35	30	30
2. ANNUAL COST OF					
– ENERGY	22	2.5	55	1.5	33
– RELAMPING	15	1.0	15	0.6	9
– SPACE CHANGES	15	0.8	12	0.3	5
PRESENT DAY VALUE			117		77
CEILING:					
1. CAPITAL COST		15	15	8	8
2. ANNUAL COST OF					
– SPACE CHANGES	15	0.3	5	0	
PRESENT DAY VALUE			20		8
FLOOR:					
1. CAPITAL COST				32	32
PRESENT DAY VALUE			0		32
TOTAL CAPITAL COST R/SQ.M		80		105	
TOTAL PRESENT DAY VALUE R/SQ.M			309		218

PRESENT DAY VALUE MULTIPLIER BASED ON :

15 YEARS	SYSTEMS LIFE
10%	INTEREST RATE
15%	ENERGY ESCALATION RATE
10%	STAFF AND MATERIAL ESCALATION RATE

The same study has been done for a building using the same type of ceiling and ceiling ambient task lights but with console air-conditioning units. Table 9 shows that the initial costs of the systems, 80 compared with 108, are considerably cheaper. However, it is difficult to achieve low running and maintenance costs with window units. Therefore the present day value of this combination of inside systems is higher — 309 as compared with 218 — even if the initial cost is lower.

The previous cost data were for the inside systems only. Table 10 gives the comparison for a total building and its systems for a good quality, console unit type system and for a structural storage access floor air-conditioning system. It shows that the initial cost for the building with console units is 260, as compared.with 295. However, when present day values are compared it shows that the cheaper capital cost solution in present day value terms is about 15 to 20% more expensive.

Table 10   Comparison for a total building and its systems.

TYPE OF INSIDE SYSTEM	FLOOR/ CEILING	ADAPTABLE CEILING	INFINITE ACCESS FLOOR & PLENUM
	ILLUMINATION	CEILING AMBIENT TASK LIGHTS	FLOOR AMBIENT & TASK LIGHTS
	H.V.A.C.	CONSOLE AIR CONDITIONING UNITS	FAN AIR TERMINALS IN OR ON FLOOR
INITIAL BASIC BUILDING COSTS		180	190
INITIAL INSIDE SYSTEMS COST		80	105
TOTAL INITIAL BUILDING COST		260	295
INSIDE SYSTEM PRESENT DAY VALUE FOR: 15 YEAR LIFE 10% INTEREST 10% ESCALATION GENERAL 15% ESCALATION ENERGY		309	218
TOTAL BUILDING PRESENT DAY COSTS		489	408

The same principle applies to the comparison between buildings with a good commercial central air-conditioning system and an access floor structural storage air-conditioning system (see Table 11). In this case, initial building costs for the two systems are more or less equal. There are some savings from reduced plantroom space in the case of the infinite access floor, but these are more than balanced out by the extra costs of the floor — so both show 310. However, if we do the comparison on the total building present day cost, we again come to the conclusion that the infinite access floor solution is approximately 15% cheaper.

Table 11 Building cost comparison.

TYPE OF INSIDE SYSTEM	FLOOR/ CEILING	ADAPTABLE CEILING	INFINITE ACCESS
	ILLUMINATION	CEILING AMBIENT TASK LIGHTS	FLOOR AMBIENT & TASK LIGHTS
	H.V.A.C.	CEILING AIR TERMINALS	FAN AIR TERMINALS IN OR ON FLOOR
INITIAL BASIC BUILDING COSTS		205	200
INITIAL INSIDE SYSTEMS COST		105	110
TOTAL INITIAL BUILDING COST		310	310
INSIDE SYSTEM PRESENT DAY VALUE FOR: 15 YEAR LIFE 10% INTEREST 10% ESCALATION GENERAL 15% ESCALATION ENERGY		289	223
TOTAL BUILDING PRESENT DAY COSTS		494	423

Chapter 13

Energy generation

J Sinnett

For the purposes of this discussion 'Energy Generation' is taken to mean 'the production of electricity', but the important side effect of 'the utilisation of the waste heat' is also considered as part of the title, as it creates the economic validity in all but very few cases. The size and type of energy generating systems under consideration are generally in the range of up to 20 MW output. There is a historical aspect to small private generating plants in the UK, which have become, with one or two notable exceptions, nearly extinct. The generation of electricity today is monopolised by the state generating boards. Some thirty years ago there were many private industrial generating plants, all coal fired, whose economics depended upon integrated heat absorbtion systems and the availability of highly skilled cheap labour to operate and maintain them. With the advent of cheap unlimited oil supplies, two things happened:

a) everybody not actually sitting on a coal field or politically constrained in some way converted to oil — this in itself helped to put a stop to small generating plants as the boilers hitherto fired for some years by coal were often unsuitable for oil, due to weaknesses in the flame tubes at grate level, and this constrained the insurance companies to derate the operating pressures, and

b) a reliable supply of cheap electricity was introduced, and also price structures and tariff systems which penalised the user unmercifully in the event that his own plant suffered a breakdown or was reduced in capacity.

This state of affairs has existed until recently, but is now coming under review in the light of present electricity prices. The state now has to pay heavily for its fuel to make electricity, as we no longer control the foreigners who own the oil-fields and for sociological reasons our own mining industry is unlikely to produce very much more output. As the generating industry is there to generate electricity, that is precisely what it has been designed to do, and what it does with great efficiency, within the limits of the technology for large generating plants. However, as the best attainable generating efficiencies are in the region of 37% and the average for the country is more in the region of 30%, then because these plants only make electricity, the remainder of the energy is thrown away.

There has been a growing interest over a number of years in district heating and also in combined heat and power schemes involving existing state power stations as well as newly installed plant by private enterprise or local authorities. There have been, and still are, considerable difficulties in promoting these aims and we still await the first major whole town plant of either kind. Under the present energy conscious climate the electricity supply authorities are starting to consider independent electricity generation schemes in a less unfavourable light. They are also collaborating with various bodies involved in specific lobbies to gain the finance and permission to utilise the waste heat from conveniently located power stations.

However, a major objector is another state body which supplies fuel — the gas supply boards. They argue that the gas distribution network will become uneconomic and they will eventually be denied the domestic heating market. There are many other interested parties involved in the arguments for and against the use of waste heat in this way.

Regardless of other arguments and points of view, the facts of the matter in fuel utilisation terms are very clear. Electricity will always be expensive due to the limitation of the thermodynamic cycle. This is based on the first and second laws of thermodynamics, which say that if you require a higher grade of energy, either electricity or work, and you wish to produce it from a heat source, then in order to produce it you have to throw away some of the heat. At 35% efficiency from a fossil fuel fired boiler plant generating electricity, for every 35 kW produced, 65 kW are rejected. A private house boiler or central heating boiler on a larger scale readily obtains operating efficiencies between 50% and 85% direct from fossil fuels, which illustrates that electricity for heating is uneconomic. These facts are illustrated very clearly in the BRE Current Paper No 56/75 "Energy Conservation : A Study of Energy Consumption in Buildings". It therefore follows that the economic generation of electrical energy is dependent upon its by-product — waste heat.

There is now a revival of interest in generating plant under the two heads —

a) central heating and power (CHP) for domestic residences, and

b) industrial plants which can utilise the availability of heating systems or process heating requirements in the factory.

Under the second head there is also the incentive of security of supply as a large population of process industries suffers seriously with product damage when power cuts occur. The state electricity supplier cannot control shutdowns due to industrial action, which have become more prevalent.

There are a number of reasons for installation of generating plant which affect the choice of equipment because of differing duty requirements and there is another series of equipment choices to match the appropriate heat requirement to electrical loadings for each application. These are briefly outlined. The reasons for the plant installation can be as follows:

1. *Stand-by plant*
   This is for security of supply when the state supplier fails. This plant is lightweight and cheap as a general rule, because the theory is that it does not have to run often and the financial trade-offs at the time of installation are indeterminate. Heat recovery is never applied in this case.

2. *Base load generating plant*
   This type of installation deals with a constant output of electricity and produces a constant output of heat. This concept has hitherto been unacceptable to the state suppliers because they do not wish to supply the variable demand component because of their own machinery load factors. The supply charge per kW in this case is therefore very high, if the supplier will even agree to a connection being made. This type of plant is the most economic from the private investment point of view because the plant can be accurately sized to a selected steady load and arrangements can be made to absorb a steady predictable supply of waste heat.

3. *Peak load generating plant*
   This type of installation deals with the situation where the demand is very variable to a large degree and its economics are determined by the savings made in reduction of maximum demand charges by state suppliers. The tariff structures of the supplier are such that the cost of total units consumed is sometimes less important than the penalty or maximum demand charge for a connected consumption level in any given thirty minutes. A typical example would be a small laundry with an installed machine load of, say, 250 kW, but an average consumption of only 100 kW. For a typical 50 hour week, if all the machines run together for several short periods in the week at 250 kW input, the penalty charge component could be almost as large as the charge for the total units.

   This type of installation may have heat recovery applied if the electrical load can be reasonably reliably predicted.

4. *Total generating plant*
   This plant depends entirely on the usage of the waste heat produced for economic viability. The nature of the electrical load and the peak/average load ratio can make this difficult to accomplish unless some heat storage can be provided to cover the discrepancies in heat production and demand which can occur from time to time. There are a number of ways of creating more favourable economic conditions apart from

Table 1 Useful energy output index for gas turbine installations – machine size **100 MW**. (Reproduced by courtesy of The District Heating Association)

Type	Fuel in nett MW	Gross CV efficiency	Elec- tricity MW	Nett CV efficiency	Heat MW	Gross CV efficiency	Nett CV efficiency	Cost index	Heat elec- tricity ratio	Z* electricity lost/heat sent out	Useful energy output index
Gas turbine + electricity	243	30.45	78.5	32.5	—	30.45	32.5	100 Oil	—	0.00	0.70
Gas turbine + electricity + boiler 127°C	243	28.8	75.0	30.8	112	72.149	77.0	127 Oil	1.49	0.16	1.20
Gas turbine + electricity + boiler 127°C	243	29.04	78.2	32.2	108.8	69.3	76.9	127 Gas	1.39	0.173	1.22
Gas turbine + electricity + boiler 95°C	243	29.04	78.2	32.2	132	77.5	86.0	127 Gas	1.68	0.142	1.32
GT + electricity + boiler + turbine 127°C	243	38.2	99.1	40.8	89	72.149	77.0	181 Oil	0.89	-0.064	1.34
GT + electricity + boiler + turbine 95°C	243	38.33	103.3	42.5	108.5	78.4	87.0	181 Gas	1.04	-0.058	1.47
GT + electricity + boiler + turbine IT 95°C	243	37.87	102.3	42.0	109.5	78.4	87.0	190 Gas	1.07	-0.040	1.46

Assume natural gas gross CV : 38.86 MJ/m³, ratio 0.902
  nett CV : 35.05 MJ/m³.

Assume oil gross CV : 19 300 BTU/lb, ratio 0.937
  nett CV : 18 100 BTU/lb.

Efficiency of best hot-water boiler: gas 82–84% gross CV
  oil 84–86% gross CV.

CHP useful energy output index of an oil-fired 36% efficient condensing station is 0.92.

*Z based on 36% efficient station.

Useful energy output index: 0.90 gas, 0.92 oil.

Note: Tables 1–3 are taken from a paper by W R H Orchard MA, MBA, C Eng, M I Mech E, presented at the District Heating Association's Second National Conference held in March 1977. Copyright DHA and author.

selection of different types of equipment. The heating or cooling systems in buildings or on process applications can be selected to improve the performance of the generating plant both thermodynamically and with respect to load factors.

An example of this approach is to increase the electrical base load of the overall system to provide more heat as a side product or to create a more stable and predictable electrical load profile. The use of electrical thermal storage heaters can help to achieve this. The use of thermal/mechanical drives instead of electrical motors for large drives can reduce peaks. There are many other considerations which can be made according to the application, but the cost of heat produced in this way is comparatively very cheap.

Across the four reasons which dictate the plant purposes is the choice of machinery, such as:

1. Diesel generators.
2. Gas turbines.
3. Steam turbines.

The appropriate selection of machinery or combination of machinery is made primarily by considerations affecting the quantity and grade of heat required.

1. *Diesel plants*

   The diesel has a very good relative thermodynamic efficiency and accordingly produces less waste heat. The heat is low grade (low temperature) but is suitable for heating systems, but rarely process use. The turn down ratio is quite good as the part load efficiency down to 33% is good.

   There are many moving parts and regular maintenance is essential. Diesel plants have a favourable first cost advantage, but higher maintenance and shorter component life than turbines.

2. *Gas turbines*

   These are relatively inefficient machines in terms of power generation (15%), but produce large amounts of high grade heat as hot gas. This gives great flexibility as the hot gas can produce either steam for heating or further power drives, for example, refrigeration compressors or different grades of water. These plants are expensive and have poor part load running characteristics. The quality of fuel is important but they can operate on gas or light oil. For economic viability there must be a large requirement for heat or power to absorb the waste energy.

3. *Steam turbines*

   These are inefficient machines although superior to gas turbines. Their principle advantage is that they are more flexible. There are several types, whose performance characteristics and flexibility vary substantially. The main types can be classified as:

Table 2 Comparison of performance of different machines for identical on steam conditions and flow from major manufacturer 100 MW series. *(Reproduced by courtesy of The District Heating Association)*

	Condition	*+ TOC or pass out*			*Back pressure*		
	Full con-dition	*Max con-dition*	*Max 127°C heat*	*Max 95°C heat*	*127°C 71°C heat*	*95°C 50°C heat*	*95°C 35°C heat*
Nett electrical MW	100	96.1	72.4	82.9	73.0	83.5	84.7
Nett heat MW	0	0	175.8	165.6	184.2	173.6	172.2
Heat/Electricity ratio	Inf	Inf	2.43	2.0	2.52	2.08	2.03
CHP UEOI	0.85	0.81	1.24	1.29	1.27	1.32	1.33
Z* electricity lost heat out	Inf	Inf	0.19	0.14	0.18	0.13	0.12
Electricity EFF gross CV oil	33.2	31.9	24.1	27.5	24.3	27.7	28.1
Electricity + heat EFF gross CV oil	33.2	31.9	82.4	82.6	85.4	85.4	85.3

Fuel in 282 MW nett, 301 MW gross.

Steam conditions 538°C 103 bar no reheat two stage hot condensing.

*Z based on 35% efficient station.

Table 3 Combined heat and power useful energy index. *(Reproduced by courtesy of The District Heating Association)*

1. Condensing set 36% M oil fired	0.92
2. Heat pump COP 2.5, 3.0, 3.5, 4.0 2.6 fuel in 1 elec out 0.9 elec to heat pump 3.0 out	0.87, 1.04, 1.21, 1.38
3. Heat pump COP 2.5, 3.0, 4.0, no transmission loss	0.96, 1.15, 1.34, 1.54
4. Small back pressure set CHP Castor Peterborough figures (17.3 × 2.39 + 60) divide by 96.1	1.05
5. Diesel combined heat and power station Aldershot (38.55 × 2.39 + 22.78) divide by 100	1.15
6. 100 MW back pressure set 127°C 71°C district heat water	1.27
7. 100 MW ITOC set 127°C 71°C district heat water	1.24 max heat 0.81 max cond
8. 100 MW back pressure set 95°C 35°C district heat water	1.33
9. Gas turbine 78 MW electrical	0.77
10. Gas turbine + boiler gas 95°C district heat water	1.32
11. Gas turbine + back pressure steam turbine combined cycle 100 MW electrical 95°C district heating water	1.46
12. 8 evaluated with 10% heat loss in distribution to give a direct comparison with 2	1.27

Note: The existing back pressure turbine 103 bar 538°C has a higher CHP useful energy index of 1.27 even including heat distribution losses than any heat pump including electrical distribution losses, unless the heat pump has a COP of 4.

Table 4 Extract from Department of Energy Bulletin "Energy Trends". Supplementary data – prices of fuels used by Industry.* *(Reproduced by courtesy of Her Majesty's Stationery Office. Crown copyright)*

		Invoiced to large industrial consumers**					Realised in new and renewed medium size contracts†	
		Coal	Heavy fuel oil	Gas oil	Gas	Electricity	Heavy fuel oil	Gas oil
			£ per ton		Pence per therm	Pence per kWh	£ per ton	
1974	1st quarter	8.5	23.6	43.2	2.64	0.835	34.7	52.0
	2nd quarter	9.1	33.3	50.6	2.78	0.854	34.2	53.5
	3rd quarter	9.8	32.8	51.8	3.03	0.924	33.8	53.1
	4th quarter	11.8	33.4	51.6	3.44	1.095	40.7	57.2
1975	1st quarter	13.3	38.8	55.0	3.65	1.179	41.4	58.4
	2nd quarter	15.1	38.2	54.2	4.15	1.178	40.9	54.6
	3rd quarter	14.9	37.5	50.7	4.43	1.249	39.7	53.2
	4th quarter	15.6	38.7	53.9	4.84	1.354	46.8	69.2

* Prices for oil products include hydrocarbon oil duty.

** Derived from the volume and value of all purchases invoiced during each quarter by a panel of about 800 large users of fuel in manufacturing industry (excluding iron and steel).

† Derived from prices reported by the ten main oil marketing companies and relate to average net prices realised on medium sized new contracts or contracts which are renewed at a changed price. The coverage may extend beyond manufacturing industry and may include, for example, larger commercial users.

Table 5 Comparison of energy costs for October 1978 issued by NIFES. *(Reproduced by courtesy of The National Industrial Fuel Efficiency Service Ltd)*

Energy form	Unit of supply	Calorific value (Btu/unit)	Average price (pence/unit)	Average price (pence/therm)	Unit supply	Calorific value (kJ/unit)	Average price (pence/unit)	Average price (£/MWh)	Price increase in past 3 months
Electricity	kWh	3413z–kWh	2.25 p/unit	65.9	kWh	3600/kWh	–	22.50	Nil
Natural gas	ft³	1035/ft³	18 p/therm	18–23	–	–	–	6.15	Nil
Fuel oil									
35 s	Gal	164 000/gal	39.31 p/gal	23.96	Litre	38 000/l	8.64 p/litre	8.17	Nil
200 s	Gal	175 000/gal	36.26 p/gal	20.72	Litre	40 570/l	7.97 p/litre	7.05	Nil
950 s	Gal	176 000/gal	33.72 p/gal	19.15	Litre	40 800/l	7.41 p/litre	6.55	Nil
3500 s	Gal	177 000/gal	32.76 p/gal	18.50	Litre	41 000/l	7.20 p/litre	6.33	Nil
Propane	Ton	21 690/lb	£106.67/ton	21.95	Tonne	50 400/kg	£105/tonne	7.48	Nil
Butane	Ton	21 340/lb	£79.24/ton	16.57	Tonne	49 600/kg	£78/tonne	5.65	Nil
Coal	Ton	12 000/lb	£29.54/ton	10.99	Tonne	27 900/kg	£29.08/tonne	3.74	Nil
Industrial coke	Ton	14 150/lb	£71.40–£72.65/ton	–	Tonne	32 900/kg	£70.27–£71.50/tonne	–	Nil
High quality blast furnace coke	Ton	14 150/lb	£72.65/ton	–	Tonne	32 900/kg	£71.50/tonne	–	Nil
Large foundry coke (S Wales origin)	Ton	14 300/lb	£79.45/ton	–	Tonne	33 300/kg	£78.20/tonne	–	Nil

October 1978 — The costs shown are approximate and apply to consumptions of 100 000 therms per annum or more. Rebates are excluded and these may be considerable in specific cases. Coal prices vary with variations in calorific value: the quoted prices apply to Daw Mill Singles delivered to the Birmingham area.

All prices except those for coke are 'as delivered' to the point of use.

a) *Condensing machine.* This machine has constant specific steam consumption/electrical production ratios at different loads. The state electricity suppliers utilise these machines in very large versions.

b) *Pass-out/condensing machine.* This machine has the ability to provide steam at any condition between the inlet conditions and atmospheric pressure, and the steam input-output can be varied independently from the electrical load.

c) *Back pressure machine.* This machine operates at a specific pressure, that is to say the machine extracts power from the steam whilst acting as a reducing valve. The consumption steam can be very high, depending upon the back pressure condition chosen; however, for large low-grade heat loads (eg, district heating applications) it is very adequate.

It can be seen that there are many alternatives or combinations of machinery available to generate energy and this has been the case for some considerable time. The establishment of the validity of these types of installation must be carried out carefully and consideration must be given to all the sub-systems within the project to establish a true financial picture for any proposal.

It is often very difficult to establish from the client precisely how financial investment is considered, and an even more difficult problem is to get certain information concerning marketing and product profitablity, which introduces assumptions and renders any analysis less precise. In recent times, now that inflation is not predictable and fuel costs are subject to political instability, cognisance of escalating fuel costs over the working life of the installation must be taken.

There are two general fears held by clients concerning (a) reliability and (b) availability of highly skilled attendance. The first fear can be designed out by correct machine selection relative to load profile characteristics and diversity studies, together with emergency allowance dependent upon the application. This, of course, with difficult load characteristics, may reduce the economic viability and would have to be assessed by the client.

The second fear is very real, but until there are installations of these types built then the people to run them will never exist. The general calibre of operating engineers in commerce and industry is very poor and there are no set standards of education or technical performance required as in the mainland of Europe or in the USA. This sort of legislation is overdue and could show savings to fuel users on such items as insurance rates and other external on costs. There would definitely be savings to be had in maintenance and spares and more efficient operation with properly trained and qualified staff.

Finally, to illustrate the advantages of generating plants with heat recovery, Table 1 indicates the variation in useful energy output index for gas turbine installations on a constant machine size of 100 MW. Table 2 illustrates the variation in overall plant efficiencies with differing types of steam turbine with constant inlet conditions and variations on the outlet conditions. Table 3 illustrates combined heat and power useful energy index for a number of existing plants. There are two heat pump examples included for comparison, although they have not been considered in this paper. Table 4 illustrates the escalating costs of fuels during 1974 and 1975 and Table 5 illustrates energy cost variation for October 1978.

Chapter 14

Energy effects of HVAC systems

T Marriott

HVAC systems are basically transportation systems, with energy as the product to be transported. If we consider that transportation takes place in two types of vehicle, one that can only freewheel (heat exchangers, pipes, ducts, etc) and one that can only climb (heat pumps), the analogy is close enough to make sense, and some of the same basic rules apply.

It is suggested that these be to:

1. Minimise the amount of product to be transported.
2. Transport it the shortest possible distance in the most efficient containers.
3. Freewheel whenever possible, and when climbing is essential keep the height to a minimum.
4. Take advantage of all free rides.

There are many more rules, or at least rules of thumb, but trying to conform to the above few is difficult enough, as will be seen as we explore them a little further.

SIZE OF PRODUCT

The nett size of the product is finally dictated by others, and in comfort air-conditioning applications is the result of the effect of lighting systems, the extent of glazing and shading, outside air requirements and the like. The only way in which the amount of energy to be carried into, out of, and all round the building can be reduced is by explanation to and education of those in the relevant disciplines. We certainly need to help them to help us, because when the building fabric, lighting, etc, is fixed, we fall into the position of a following trade, and have to mop up the results.

It is at this conceptual stage, therefore, that priorities and aims must be sorted out, and the entire design team must understand whether we are attempting to design to reduce running cost, to minimise total building owning and operating cost over some defined time scale, to help keep the national energy consumption down, or to fulfil some other aim. Without some common understanding decisions will be taken with no idea of how far from (or near to) the optimum they are, or just what opportunity cost is being incurred. It is unreal to pretend that buildings will be designed purely with minimum energy in mind — what is important is to know how far any decision takes us from the optimum energy solution.

The gross size of the product is determined largely by the control systems applied to the HVAC systems. It is quite common that when less cooling, for instance, is required, what is provided is full cooling plus partial heating. Common dew point plants with terminal reheat are almost bound to fall into this category at some time of the year. Induction unit systems suffer from the same problem, where in winter, for example, the primary air is made hot enough for the room with the worst heat loss, and all the other rooms in the same zone receive equally hot air which has to be cooled in the terminal. A central cool air supply fighting an overscheduled perimeter heating system is another good candidate in the inefficiency stakes, as is our old favourite, overdriven radiators with seized valves, with the occupants opening windows for temperature control. Obviously we should devote as much attention to avoiding that sort of load magnification as to talking our colleagues out of excessive glazing, tungsten lamps, cold bridges or non-draughtproof facades.

Naturally, aspects such as load factor are also important. In electrical supply and distribution the effects of a poor load factor on the efficiency of use of capital plant and distribution equipment has long been recognised. It is particularly important when there is no buffer between what is generated and what is used. To use the transport analogy, when a factory has no warehouse or other storage, and the rate of work in the factory is variable, the scheduling of vehicles bringing raw materials and taking away finished goods is difficult and critical. The utilisation of mechanical systems plant operates in exactly this way, and should be so regarded. Many of the load changes in a building are not within the control of the designer, but some simple rules are worth keeping in mind, as they do help to ease the pain. These are:

1. Surface admittance and building weight make a lot of difference to the inertia of the building and its ability to smooth out load variations.

2. Similarly, flywheels can sometimes be found on the water side, though the time constant of anything less than a swimming pool is likely to be shorter than that of the air side.

3. Examine the likely time constants of the building and systems to assess the dynamic response required from the controls. True dead-beat control is important for economy; too many systems spend too much of their time in a gently hunting state.

4. Look at anything in the diurnal, or even weekly cycles which may help and try to bring into phase with the building needs.

5. Allow as much swing in room condition as possible.

EFFICIENCY OF TRANSPORT

Minimising transport distance is a function of planning the geography of the building and the systems sensibly, and organising a sensible distribution harness. Choosing the most efficient containers has long been a matter of debate. Water is a much better fluid than air for the transport of energy, and this has probably been a major factor in the development of various air/water systems. Air is still necessary, because it is what the occupants breathe, and therefore has to be collected from the outside, distributed in controllable quantities to every part of the building and then collected again and discharged.

The usual control of air/water systems is that the air is treated in bulk on some sort of schedule, and terminals are supplied with water for the final control. The usual result is that at non-design states the terminal is offsetting a proportion of the central treatment. The gross load is then considerably in excess of the nett load.

Variable air volume systems were invented to overcome this sort of trouble, and to keep control simple, air is always supplied at less than room temperature.

This is ideal for internal zones, which always have a nett heat gain and therefore need cooling, but not so good for perimeters which need further local treatment. For large buildings it has significant advantages, not least of which is that it allows for diversity of load right back to the air handling equipment, so that ducts and fans can be selected for less than the sum of the individual loads. The power absorbed by a fan is proportional to the air flow cubed, so even a 15% reduction in load is worth having, as it reduces the absorbed power by almost 40%. However, with the mechanisms available to us at present it is disappointing to see just how much the motor's power factor decays as load is reduced, resulting in a depressingly small reduction in motor current. This is due to the decreasing efficiency of electric motors with decreasing load.

It has been suggested that a figure of merit should be calculated for air-conditioning systems. This would be the maximum building load divided by plant power consumption, and would provide a comparison in energy consumption terms of various systems. It is a very useful concept, but it needs to be extended to cover non-design running conditions to give a true picture.

FREEWHEELING AND CLIMBING

This is just realising the fact that heat will flow by itself from a warmer body to a cooler one and has to be pumped from a cooler body to a warmer one — a sort of thermal gravity.

It is obvious, therefore, that when a building is to be held at a temperature below that of the outside air, some heat pumping must take place. The heat produced in the building must be removed by something colder than the room air, and must be pumped to a temperature above that of the outside air in order to reject the heat into the atmosphere. The power required to pump a given quantity increases with the temperature difference across which the pumping takes place. It is advantageous, therefore, to cool with as warm a fluid as possible, and heat with as cool a fluid as possible, to keep the power consumption of the heat pump to a minimum.

Where the heat rejected from the heat pump is used for heating, as opposed to being removed in a cooling tower or something similar, the choice of system temperature is important. It should be as low as possible to save energy in the heat pump, losses in distribution pipework, etc, and as high as possible to keep the final heat-exchangers to room air small.

In some parts of the process there will be temperature gradients which are potentially useful, and we should take advantage of them wherever possible. Some of these, such as extracting air from a room over the light fittings, have been used so widely and for so long that they pass almost without comment. Probably, if they were first used today, they would be heralded as great energy savers. Such is the effect of fashion.

ACCEPT FREE RIDES

This is meant to cover the probability that the building is going to contain something that can add to the elegance of our schemes. All too often designers treat the client's equipment as an unpleasant excrescence on their schemes, and not as something to be integrated into the overall balance of the building.

The one thing to watch here is that, to take our transport analogy, designers are asking a stranger for a long lift, and ask him where he is going before letting him know where they want to go. It is vital, therefore, to be on the look-out for false information. In some cases the client may wish to impress designers with the power of his organisation, or may just be too enthusiastic. He may be confused by the designer's need to know the maximum load he may put on the building systems (so that things can be sized adequately) and the minimum load he can reasonably guarantee (so that designers can take advantage of it with some confidence). The nature of his operation may genuinely change, resulting in a different balance.

For all these reasons, the client's equipment load has to be treated with caution where designers allow it as an advantage to their scheme, while remaining somewhat pessimistic where it is a disadvantage. In other words, accept a lift if you are fairly sure where the driver is going, but try to make certain that you will not be hurt if you have to bale out en route.

Chapter 15

Energy audits

D Gurney

INTRODUCTION

Until the 1970s energy was considered to be both inexhaustible and expendable, assumptions seemingly verified by the relatively low cost of most forms of energy. The problems fostered by such attitudes are exemplified in the design of commercial buildings from that era. Typically, buildings were designed and constructed primarily with initial costs in mind, and the result has been the creation of a vast collection of commercial buildings which — by today's standards — utilise excessive amounts of energy.

It is fairly safe to say that buildings being designed now, and those to be designed in the future, will utilise many of the new techniques and systems which can lead to maximised energy efficiency. The rate of building replacement is very slow, however, and as a result the majority of existing commercial and industrial buildings for many years to come will be those which were originally *not* designed with energy conservation in mind.

Two methods for achievement of energy conservation have been advanced:

1. The implementation of specific end use restrictions in buildings, for example, adjustment of thermostats to specified levels, reduction of lighting levels to reduce consumption, but the end of use restriction method has numerous drawbacks. Key among them is the fact that the extent to which a system is used has no bearing on its efficiency.

2. The energy conservation approach (called 'Total Energy Management' in the USA, or TEM for short). In short, this considers every building as a unique, complex system, and to conserve energy we must first understand how the building consumes its energy, how users' needs are met, and how the systems interrelate, particularly with regard to the external environment. The techniques applied in these studies are now commonly known as 'Energy Audit'.

The procedures and techniques involved in Energy Audit are not new. What is new is the manner in which they are addressed and the methodology suggested for the effective implementation. Most important is the realisation that this approach will enable building owners, managers and designers to realise significant energy and energy cost savings without having to make any significant changes in the working environment.

It is perhaps pertinent to state the obvious: that unless we are expert in energy conservation techniques as applied to existing buildings, our new designs will not necessarily reflect this knowledge.

ENERGY CONSUMPTION IN BUILDINGS

Energy use in buildings is determined, basically, by climatic conditions of the area in which they are located and the working environment required. Neither of these factors is capable of significant modification: nothing can be done about the climate, and very little about altering tenants' requirements.

The efficiency of energy use is determined by three basic systems which comprise any functioning building:

a) energised systems such as those required for heating, cooling, lighting, ventilation and lifts;

b) non-energised systems such as floors, ceilings, walls, roof, windows;

c) human systems comprising maintenance, operating and management personnel.

Each of these three systems is capable of modifications which can lead to a significant saving in energy. Effective energy conservation requires that the entire pattern of energy consumption be analysed so that changes can be made in the full light of the interrelationships which exist. For example, energy conservation in terms of heating and cooling is not simply a matter of adding or removing heat from the inside air to achieve a given temperature. To consider this in isolation would be to ignore what really is involved, namely, compensating for heat losses and gains in the building which occur simultaneously. The factors which influence heat gain and heat loss are as follows:

1. *Infiltration.* This involves the passage of outside air into the building through apertures such as cracks around windows and door jambs, doors left open, etc. The amount of infiltration depends on the impact of the wind and on the integrity of construction.

2. *Transmission.* This relates to the amount of heat transmitted into the building or lost from it through the various components of the building envelope, including exterior walls, windows, roof, doors, floor, etc.

3. *Ventilation*, which makes a contribution to heat gain and loss depending upon the season involved.

4. *Lighting.* This contributes to a building's heat gain in direct proportion to the power of the lamps. This heat gain is beneficial during the heating season but during the hot weather the cooling system must compensate for the heat from the lights.

5. *Solar heat.* Like the heat of light, solar heat contributes to heat gain throughout the year.

6. *Equipment.* Virtually all powered equipment, including business machines, computers and mechanical services equipment, contribute to heat gain because their motors or other elements generate heat.

7. *Occupants*, who create a significant heat gain within the building.

Evaluation of the balance between the coincident heat gains and losses for various climatic conditions enables the building services engineers to evaluate the energy requirements of the building. At one time many of the internal heat gains (lighting, equipment and occupants) were viewed as a bonus and not taken into account in the calculation. This often led to overheating, especially when the temperature control systems were very simple and unresponsive.

COMMITMENT AND CO-OPERATION

In order to proceed with an energy audit it is necessary to obtain the commitment and co-operation of the various parties involved. These include:

a) the building owner and his accountant,

b) the building manager and purchasing manager,

c) maintenance and operating personnel,

d) those employed in the building.

The above applies to an existing building. In the case where an audit is concerned with a future building at present under design, then of course the commitment and co-operation of all members of the design team is just as essential. This applies particularly to the quantity surveyor, whose concern is for the first or capital cost and who in the past has taken little or no interest in the operating costs. This must now change.

THE ENERGY AUDIT SURVEY

The energy survey is essential to the successful audit but it is not an easy task to undertake. It is particularly difficult for the building operator or industrial works manager who for so long has been charged with maintaining his services at optimum operating efficiency on the assumption that his main resources of energy, labour and finance were readily available and cheap. In the new situation he has to look both forward and backwards and attempt to balance these in an unbiased manner which has considerable bias on his operating performance. For this reason an external energy consultant is now frequently employed because he can ask the unpalatable question and he may recognise that the way in which a specific function is being conducted could be altered to achieve significant savings. Likewise, he may note that maintenance is not being performed as well as it should be and can recommend improved procedures. For this reason the energy consultant is often involved also in a management consultancy role. It will be seen, therefore, that this service is not likely to be cheap, since it must involve a senior and highly experienced engineer.

The energy audit survey must be based on the record: architectural, mechanical and electrical design drawings and specifications. If such drawings are not available, it may be necessary to develop single line diagrams of the existing services and systems. The surveyor must also be given access to written maintenance and operating procedures and manuals and he should have available all fuel, electricity and water charges, preferably as invoices over at least the preceding year.

The items which require investigation and analysis are outlined in the following section and in the appendices. Just a quick glance will indicate that some of the most critical areas include airtightness of the building, the thermal transmission characteristics of the fabric, services systems operation and controls, and occupier procedures, including maintenance.

The survey should result in a report which details each and every fault of the energised, non-energised and human systems as they relate to excessive energy consumption. It should also contain a list of alternative recommendations which can be undertaken to remedy each fault in the light of the building interrelationships, and an estimate of the cost of such modifications.

Also included in the appendices is a selection of survey charts drawn from various sources. It is strongly recommended, however, that the reader prepares his own charts.

GUIDELINES FOR ENERGY CONSERVATION

The scope of such guidelines is necessarily broad to cover the many alternative types of building and systems, and not all are applicable in every case. It is useful to list the alternatives under two categories: *those of minimal expense* and *those requiring significant expense.*

There are many published energy audit check-lists, and a representative sample is included in the reference section to this chapter. The reader is particularly recommended to consult the UK Department of Energy Industrial Energy Efficiency booklets (references 1–4), and also reference 5, from which has been drawn Appendix B, being a part of the US Federal Energy Administration recommended guidelines for energy conservation. This is a very long section covering extremely detailed assessment of the building services systems, but the part selected for inclusion in Appendix B deals primarily with building planning, fabric characteristics, infiltration and ventilation.

IMPLEMENTATION OF THE PROPOSALS

Having undertaken the energy audit survey and prepared the technical report it will be necessary, in most instances, to prepare a financial statement of capital and operating costs, the prime objective of which will be to demonstrate the pay-back which can be achieved in terms of covering the capital costs invested within a comparatively short term of years using the cost savings to do this. This exercise will be essential to convincing the building owner that he should undertake the energy conservation programme.

It is also essential to be aware of the fact that the capital cost investment is aimed at savings to be made in the future wherein inflation will have a considerable bearing on the result. An effective method of presenting such information is to undertake a simple analysis of the capital cost and fuel costs implicit in not implementing energy conservation procedures for a period of, say, five years.

AN ENERGY AUDIT CONSULTANCY

This brings us to the question as to whether a firm should become involved in the energy audit field. Quite apart from the many advantages which can be put forward for doing this, including that of making new client contacts, there is one further aspect which the author feels to be important.

Such meetings as the Department of Energy's meetings of industrial energy managers and the National meetings held at Birmingham in 1977 and 1978 are always well attended, with up to 40 energy managers in regional meetings and in excess of 750 on each of the two days at Birmingham in 1978. The papers presented and the discussions have always been extremely practical and positive in their direction, and in addition the technical press is full of working examples of savings achieved.

Thus, it seems that we are facing a situation in which design team managers and design engineers could well be less well-informed on what is achievable in practice than will be the rank and file personnel among its clients. It would be sad to find that new buildings which it has designed are less economic to run in energy terms than the much older buildings which industrial and commercial clients had modified by their own efforts, and that its clients could see that this is so, even before the new designs are built.

There is a need, therefore, to ensure that project managers and building-design engineers are suitably informed and experienced (trained if need be) and up to date in the techniques and procedures of energy audit and design for energy economy in use.

The author has already proposed a solution to this problem, and within the Ove Arup Organisation positive action is being taken to implement this proposal. In one of the appendices to this chapter an outline is included which was in fact prepared in 1973 before the OPEC oil price increases created the new interest in energy conservation. In this outline brief objectives and a method for the setting up of an energy audit consultancy are set out. It is still relevant after six years, although some alterations are probably necessary to match current conditions. It is included here as a suggested starting point for those who may wish to get involved.

CONCLUSION

This chapter only scratches at the surface of what is a very wide and deep subject, and those who are interested in further information are recommended to read the documentation listed in the references.

Although the common consensus is that 'energy crisis' is perhaps too emotive a statement and that 'energy shortage' is a better view to take, there is no doubt that problems related to energy costs and availability are here to stay and will grow in importance as time passes. Building designers are, it is believed, only at the beginning of a process which will alter very considerably the buildings which they design and the industry of which they are a part, and the author also believes that they should take a very serious look at how they audit the energy to be used by their clients.

REFERENCES

1 Department of Energy, "Energy Saving in Industry", 1975.
2 Department of Energy, "Energy Audits", Fuel Efficiency Booklet 1, 1976.
3 Department of Energy, "Energy Audits 2", Fuel Efficiency Booklet 11, 1978.
4 Department of Energy, "Energy Management and Good Lighting Practice", Fuel Efficiency Booklet 12, 1977.
5 United States Federal Energy Administration, "A Practical Handbook on Energy Conservation and Management", no date.
6 Property Services Agency, "Energy Saving Action Chart for Existing Buildings", HMSO, 1976.
7 Esso Petroleum Co Ltd, "Energy Saving in Industry", Esso Petroleum, 1975.
8 Cheshire County Council, Energy Conservation Unit, "Energy Conservation in Cheshire", Cheshire County Council, 1979.
9 Gurney, J D, "Conservation and Recovery of Heat in Building Design", Building, 230 (35), pp 65–66, 1976.
10 Chartered Institution of Building Services, "Building Energy Code Part 3: Guidance towards Energy Conserving Operation of Buildings and Services", 1979.
11 American Society of Heating, Refrigerating and Air-conditioning Engineers Inc, "Energy Conservation in New Building Design", Standard 90–75, 1975.
12 Arthur D Little, Inc, Consultants, "An Impact Assessment of ASHRAE Standard 90–75". See also ASHRAE Journal April, May, June 1976 for extracts:
 a) Lentz, C, ASHRAE Standard 90–75: Impact on Building Energy Usage and Economies. ASHRAE Journal, 18 (4), pp 23–28, 1976.
 b) Lentz, C, ASHRAE Standard 90–75: Potential Impact on National Energy Demand. ASHRAE Journal, 18 (5), pp 23–27, 1976.

    c) Lentz, C, ASHRAE Standard 90–75: Economic Impact on Selected Industries and the Design Profession. ASHRAE Journal, 18 (6), pp 33–38, 1976.

13 Blossom, J S, and Bahnfleth, D R, "Upgrading Plant Ventilation: a Study", Heating/Piping/Air Conditioning, 48 (4), pp 81–87, 1976.

14 Kirkwood, R C, "Energy Conservation in Industrial and Commercial Buildings", paper presented at the Royal Institute of British Architects, National Energy Management Conference, Birmingham, 1978.

15 Braham, G D, "Conservation and Management of Energy", paper presented to the Institute of Baths Management, 1975.

16 Building Research Establishment, "Energy Consumption and Conservation in Buildings", Digest 191, 1976.

17 Rudoy, W, and Duran, F, "Effect of Building Envelope Parameters on Annual Heating/Cooling Load", ASHRAE Journal, 17 (7), pp 19–25, 1975.

18 Patterson, N R, and Alwyn, J B, "The Impact of Standard 90–75 on High Rise Office Building Energy and Economics", Heating/Piping/Air Conditioning, 48 (1), pp 38–44, 1976.

APPENDIX A. TABULATED DATA

Appendix A1

(Extract from Energy Paper No 12 of the Department of Energy: "Advisory Council on Energy Conservation: Paper 3, Energy Prospects". Crown copyright)

...we suggest that oil supplies are likely to become constrained within the period 1985 to 1995. For the low economic growth case oil scarcity might commence within the period 1990 to 2000. The range of uncertainty shown for oil supply in Figures 5 and 6 arises in the short and medium term from uncertainty about the level of exports that would be permitted by the major oil exporting countries. In the medium and long term the range follows from uncertainty about the rate of discovery of new oil reserves coupled with an assumption that the average reserves to production ratio world-wide should exceed 15. The higher and lower bounds on the shaded regions for potential oil supply in the longer term correspond approximately to annual discovery rates of 18 and 12 billion barrels respectively. This may be compared with the current annual use of about 17 billion barrels and our projected peak in possible oil supply around 1990 which is about 27 billion barrels a year.

Our estimates for future supplies of other fuels are indicated in Figures 8 and 9. The limits on the growth of coal arise from lead times for the development of world trade in coal, and limits that may be set on the export of coal from the United States, the USSR and China. Gas supplies are assumed to be resource limited in a manner similar to oil, but with further constraints arising from transport problems. The growth of nuclear power is assumed to be delayed, owing to the present excess capacity in the electricity industry. Its rate of growth is also assumed to be inhibited by doubts about the availability of uranium, which may remain as a constraint until commercial fast breeder reactors are fully operational, perhaps late in the 1990s.

Additional constraints on the growth of nuclear power include the capital costs of large nuclear programmes and social or political concern that may limit or delay nuclear development in some countries...

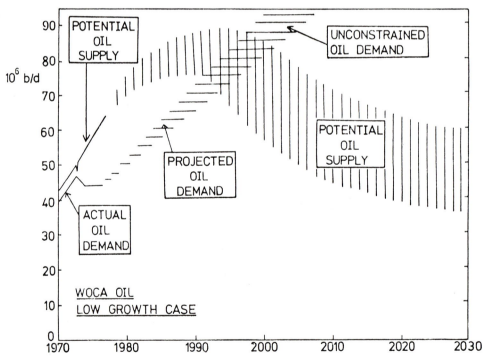

Fig 7   World (WOCA) oil supply and demand projections (low case)

Do you know how much those leaks cost you?

COMPRESSED AIR — 100 P.S.I. (7 BAR)

Size of leak	Cubic feet (free air) wasted per month (200 hours)	Cost per month (See below)
1/32"	24,000	£2.78
1/16"	78,000	£9.04
18"	278,400	£32.28
1/4"	821,759	£95.32
3/8"	1,775,552	£205.96

Notes
1. Costs based on 5.8p to raise 1000 cu. ft. plus additional power required to make good leakage deficits.
2. Based on electricity cost average of 2.1p per kw. Smaller concerns tariff may be as high as 2.56p; continuous progress industry tariff as low as 1.8p.
3. Without leaks energy cost of compressing 1000 of an on 48 hour week amount to £8376 per year. Even a 10% saving is therefore important.
4. Electricity costs have risen by approx 145% since 1976.

STEAM — CONTINUOUS PROCESS

165P.S.I. (11 BAR)		100 P.S.I. (7 BAR)	
Pounds force wasted per month	Cost per month at £2.20 per 1000 lbs**	Pounds force wasted per month	Cost per month at £2.20 per 1000 lbs**
4,752	£10.45	3,250	£7.15
19,130	£42.08	13,00	£28.60
74,050	£162.91	52,500	£115.50
302,458	£665.40	210,000	£462.00
678,985	£1493.76	470,000	£1034.00

Notes
1. Costs based on heavy fuel oil at avge industrial price of 6.16p per litre and avge boiler efficiency of 80% (many reduce sooner) exclusive of all other overhead costs.
2. A 1/8" steam leak can lose energy equivalent to approx 16 tons of coal per year. 540 million Btu per year at 1000 PSI.
3. A 1/4" leak caused by a malfunctioning steam trap can waste 1.75 million Btu per month. Approx £450 wasted.

WATER (METERED SUPPLY) 60 P.S.I. (4 BAR)

Gallons wasted per month	Cost at 66p avge per 100 gal
4,994	£3.29
20,013	£13.20
77,802	£51.35
317.694	£209.60
752,799	£496.81

Notes
Cost vary between regional authorities e.g. lowest — Thames 56.5p, highest — S. West 80.2p.
Water costs will rise by average of 14% in 1978/79.

Average figures based on Government and industry related sources (D.O.E., U.S. Dept of Commerce, British Compressed Air Society, ESSO Petroleum Ltd, Electricity Council, Regional Water Authorities)

(Reproduced by courtesy of Furmanite Engineering Limited)

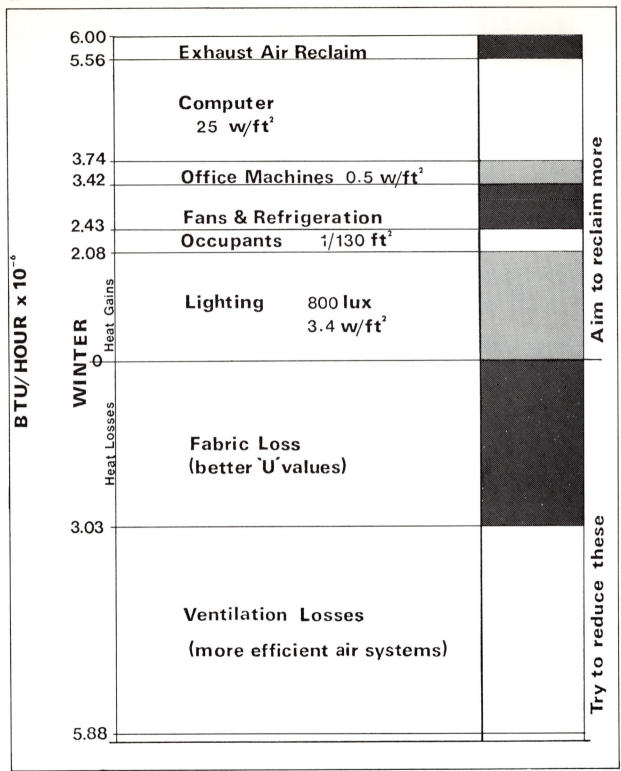

Heat gains and losses in an office building under typical UK winter conditions.

APPENDIX B. EXTRACT FROM "GUIDELINES FOR ENERGY CONSERVATION (REFERENCE 5)" *(By courtesy of the US National Technical Information Service)*

This section identifies and discusses numerous different ways in which office buildings and retail stores can be made more energy efficient. Guidance presented can be applied during the survey to establish what specifically should be looked at, as well as in determination of which energy conservation opportunities should be pursued, and how. The scope of these guidelines is necessarily broad to provide discussion of the many alternative types of sub-systems which could exist in any given building. As a result, some of the items discussed — especially in the area of heating and cooling equipment — will not be applicable to your particular building. Except as noted, however, most discussion is applicable to both retail stores and office buildings.

The specific alternatives discussed are broken into two categories: minimal expense and significant expense. In certain cases some of the minimal expense items will be checkmarked ($\sqrt{}$). This indicates that they involve virtually no cost at all; have little or no impact on occupant comfort and business productivity, and can be implemented immediately to reduce wastage of energy. Such steps would include, for example, lowering thermostats set at abnormally high levels during the heating season; being sure that lights, equipment, machinery and appliances are turned off when an area is not in use; keeping heat exchanger surfaces and filters clean; stopping leakage of conditioned air and water, steam condensate, fuel oil and gas, etc. In all cases, such actions — in fact, all energy conservation actions — should be undertaken in a manner consistent with comfort, production, process and other requirements of building occupants.

It must be borne in mind that the cost categorization has been supplied primarily for reference purposes. While cost is a major factor in establishing which alternatives to pursue, one also must consider the benefits involved. Moreover, cost factors vary from building to building, depending on the extent and capabilities of in-house personnel, accounting methods, and so on.

Before determining exactly which alternatives to pursue, it is suggested strongly that you first identify those alternatives applicable to your building. Since it is unlikely that you will be able to pursue all of them during the first or even second year of your TEM programme, it probably will be most effective to determine which group of alternatives — in the aggregate — will fall within budget and yield maximum benefits.

As a last note, recognize that the following discussion addresses itself primarily to components of various systems. This approach has been utilized to reduce needless repetition in that two entirely different systems may have many components in common, such as fans, motors, etc. Nonetheless, some repetition does exist where required for better understanding of material covered. It is advised, therefore, that the index be utilized to establish all places where a given component or other subject is discussed.

Interior space utilization

The way in which interior building space is utilized by occupants can have a pronounced effect on energy utilization as well as on business productivity. Some of the potentially beneficial modifications which should be considered include:

Work methods improvements
Minimal expense:
  $\sqrt{}$a) Consider combining operations and work elements, changing sequence of operations, simplifying necessary operations, eliminating unnecessary work, and establishing preferred work methods. In this way, for example, trips between floors via the elevator can be reduced if several routine tasks can be undertaken while on the floor involved, instead of making two or three trips to perform each task separately.

Equipment and materials relocation
Minimal expense:
  $\sqrt{}$a) Consider locating all computer and computer-type equipment which requires close temperature and humidity control in a common space to be served by a common system.

  $\sqrt{}$b) Where practical, place all heat-producing equipment such as duplicating machines in one area to enable easier control for heating and cooling purposes.

  $\sqrt{}$c) Locate wall hangings, displays and furniture away from supply and return air grilles and registers to prevent obstruction of air flow.

Significant expense:

√d) Consider removing partitions to create an open space effect which will permit much freer movement of air and a reduction in lighting requirements.

Other
Minimal expense:

√a) Close off unused areas and rooms. Where possible, be certain that blinds or other shading devices are drawn, registers closed, etc.

√b) If possible, have persons working after hours work in proximity to one another to lessen lighting and HVAC (heating, ventilating and air-conditioning) requirements.

√c) When repainting, consider use of light wall surfaces to reflect both heat and light.

√d) If possible, place in proximity to one another those persons whose tasks require similar lighting levels.

Infiltration

Unwanted outside air infiltrates into a building through inadvertent openings in the building envelope, open doors, etc. Since outdoor air, regardless of source, must be heated or cooled (and sometimes humidified and/or dehumidified), infiltration imposes a significant load on the heating and cooling system, increasing total energy consumption. The problem of infiltration is appreciably worse in tall buildings, which usually must contend with 'stack effect'. This occurs because of the difference in density between warm indoor air and cold outdoor air. As indoor air is heated it becomes lighter and tends to rise, eventually leaving the building through upper floors. The hot air is replaced by a continual flow of cold air entering the building through any available opening. The upflow of air increases with the height of the structure.

The following guidelines should be observed to inspect for sources of and reduce infiltration, regardless of building height or configuration.

Windows and skylights
Minimal expense:

√a) Replace broken or cracked window panes.

√b) Replace worn or broken weatherstripping around operable windows. If possible, install weatherstripping where none was installed previously.

√c) Weatherstrip operable sash if crack is evident.

√d) Caulk around window frames (exterior and interior) if cracks are evident.

√e) Rehang misaligned windows.

√f) Be certain that all operable windows have sealing gaskets and cam latches that are in proper working order.

√g) Consider posting a small sign next to each operable window instructing occupants not to open window while the building is being heated or cooled.

Doors
Minimal expense:

√a) Replace any worn or broken weatherstripping. Install weatherstripping where none has been installed previously.

√b) Rehang misaligned doors.

√c) Caulk around door frames.

√d) Inspect all automatic door closers to ensure they are functioning properly. Consider adjustment to enable faster closing.

√e) Inspect gasketing on garage and other overhead doors. Repair, replace or install as necessary.

√f) Consider placing a small sign next to each door leading to the exterior or unconditioned spaces advising occupants to keep door closed at all times when not in use.

√g) Consider installing signs on exterior walls near delivery doors providing instructions to delivery personnel on operation of doors.

√h) Establish rules for all building personnel regarding opening and closing of doors, directing them to keep them closed whenever possible.

i) Consider installing automatic door closers on all doors leading to the exterior or unconditioned spaces.

j) If the building has a garage but does not have a garage door, consider installing one, preferably motorized to enable easier opening and closing.

k) Consider use of a card-, key- or radio frequency-operated garage door which stays closed at all times except when in use.

Significant expense:

l) Consider making delivery entrances smaller. The larger the opening, the more air that infiltrates when doors are open.

m) Consider using an expandable enclosure for delivery ports. It reduces infiltration when in use because it can be adjusted to meet the back of a truck reducing substantially the amount of air which otherwise would infiltrate.

n) Consider installation of an air curtain, especially in delivery areas. The device prevents penetration of unconditioned air by forcing a layer of air of predetermined thickness and velocity over the entire entrance opening. (An expert in the field should be consulted before obtaining such a device, especially when high-rise structures are involved. The degree of stack effect, among other things, determines its usability.)

o) Consider installation of a vestibule for the front entrance of a building, where practical. It should be fitted with self-closing weatherstripped doors. It is critical that sufficient distance between doors is provided.

p) Consider utilizing revolving doors for the front entrance. Studies have shown that such devices allow far less air to infiltrate with each entrance or exit. Use of revolving doors in both elements of a vestibule is even more effective. If high peak traffic is involved, swinging doors can be used to supplement revolving doors.

Ventilation

Ventilation has significant impact on a building's total energy consumption. Each cubic foot of air brought into the building must be either heated or cooled and, in some cases, humidified and/or dehumidified. It is generally agreed that many building codes demand an amount of ventilation in excess of what actually is required to provide for the safety and comfort of building occupants. Because many building code officials also recognize this, they often are willing to allow changes to ventilation systems which will drop CFM rates below those nominally required, providing that such changes are not irreversible. Of course, the approval of code authorities must be obtained before you undertake any changes which will result in violation of the applicable codes as written. Guidelines for action are as follows:

Minimal expense:

a) Limit outdoor air to the minimum required to balance the exhaust requirements and maintain a slight positive pressure to retard infiltration-caused heat losses and heat gains.

√b) Inspect all outside air dampers. They should be as airtight as possible when closed. Check operation of position indicators for accuracy. Install, repair, or replace position indicators as needed.

c) Reduce or eliminate the need for using outdoor air for odor control by installing chemical or activated charcoal odor-absorbing devices.

√d) Inspect filters carefully. If necessary, create a filter replacement schedule. Utilize high-efficiency, low-cost filters.

√e) Reduce exhaust air quantities as practical.

f) Close outdoor air dampers during the first and last hour of occupancy when the air must be heated or cooled.

√g) Establish a ventilation operation schedule so exhaust system operates only when it is needed most.

h) Use exhaust hoods in food preparation areas only while cooking operations are underway. Add control dampers or gravity damper to keep the air path in the exhaust duct closed when fan is not operating.

i) If a food preparation area exhaust hood is oversized, adjust it so no more air than necessary is exhausted. This can be done easily by blocking off a portion of the hood, or reducing fan speed, or lowering hood, or by utilizing a combination of these techniques in compliance with applicable health regulations.

j) Consider cutting off direct outdoor air supply to toilet rooms and other potentially "odorous" areas. Permit air from other areas to migrate into such areas through door grilles and be exhausted.

k) If possible, concentrate smoking areas together so they can be served by one exhaust system.

l) Consider adding controls to shut down the ventilation system whenever the building is closed for an extended period of time, as during the evening, weekends, etc., except when the economizer cycle is in use.

m) Reduce volume of toilet exhausts in buildings which have multiple toilet exhaust fans having a total fan capacity in excess of outside air requirements. This can be done by wiring a fan interlock into toilet room lights through a timed relay, so the fan is activated only when lights are on. An administrative request plus signs to the effect that lights should be turned off when the room is not in use will help ensure that lights (and thus the fan) are off when the room is not being used. Another method involves dampering down air volume so only that amount of air required by code is removed.

n) If a large occupant load is involved, consider installing remotely adjustable outside air dampers. These permit outside air volume to be adjusted in approximate proportion to current occupancy.

o) Install baffles to prevent wind from blowing directly into an outdoor air intake.

p) Supply ventilated air to parking garages to levels indicated by CO_2 monitoring system.

Significant expense:

q) Consider installing economiser enthalpy controls to air handling units in offices to minimise cooling energy required by using proper amounts of outdoor and return air to permit "free cooling" by outside air when possible.

r) When more than 10 000 CFM is involved, and when building configuration permits, consider installation of heat recovery devices such as a rotary heat exchanger. For some climatic conditions an "enthalpy wheel", which permits recovery of some 75% of outdoor heat load during both heating and cooling cycles, will be feasible.

Transmission

Transmission losses generally can be reduced either through addition of insulation or modification of glazing characteristics.

Insulation
It is advised strongly that expert technical assistance be obtained before undertaking any insulating to help ensure that the proper type and correct amount are installed, that cost effectiveness will result, and that any potential problems — such as moisture condensation — can be avoided.

Minimal expense:

a) Where roof insulation is not practical, consider insulating the top floor ceiling. This can be done easily with blown insulation. In most cases, ceiling insulation will also require a vapor barrier placed on the warm side of the ceiling — if not integral with the insulation — to prevent structural damage caused by rot, corrosion or expansion of freezing water.

b) If remodelling or modernisation is contemplated, consider adding insulation to all exterior walls as well as those which separate conditioned and non-conditioned spaces. As a general guide, the U value for the gross external walls (including windows and doors) of 0.35—.40 BTU/h ft² °F is considered to be an attainable minimum goal.

c) Add or improve insulation under floors over garages or other conditioned areas.

Significant expense:

d) Consider adding roof deck insulation, especially if your building is 20 years old or older. Assuming that the roof/ceiling sandwich is not used as a return air plenum, a thermal transmission value (U-value) of 0.08–0.10 BTU/h ft² °F is considered to be an attainable goal through roof/ceiling sandwich.

Glazing

In all cases, preferential treatment should be given to those windows most exposed to direct sunlight or high levels of reflected sunlight.

Minimal expense:

√a) Inspect condition of indoor shading devices such as drapes and blinds, which can reduce heat gain as much as 50%. Keep indoor shading devices clean and in good repair.

√b) During the heating season, close all interior shading devices before leaving space to reduce night-time heat losses.

c) Use opaque or translucent insulating materials to block off and thermally seal all unused windows.

Significant expense:

d) Consider adding reflective and/or heat-absorbing film to glazing to reduce solar heat gains by as much as 80%. Do be aware that such films will reduce substantially the benefits of natural lighting.

e) Consider adding reflective materials to the window side of draperies to reflect solar heat when draperies are drawn.

f) Install indoor shading devices where none now exist, even if exterior shading devices are used. They should be light-coloured and opaque.

g) Consider installation of outdoor shading devices, such as sunshades, which reflect solar heat before it has a chance to enter the building, and which dissipate heat outdoors rather than indoors. Adjustable sunshades enable entrance of warming rays during the heating season.

h) Consider installation of storm windows if practical.

i) Consider reglazing with double or triple glazing, or with heat absorbing and/or reflective glazing materials.

Exterior surfaces

Minimal expense:

√a) Caulk, gasket or otherwise weatherstrip all exterior joints, such as those between wall and foundation or wall and roof, and between wall panels.

√b) Caulk, gasket or otherwise weatherstrip all openings, such as those provided for entrance of electrical conduits, piping, through-the-wall cooling and other units, outside air louvers, etc.

c) Where practical, cover all window and through-the-wall cooling units when not in use. Specially designed covers can be obtained at relatively low cost.

d) Repaint or clean exterior finish to improve reflective characteristics.

e) Repaint or resurface roof to make it more reflective.

Elevator shafts

Minimal expense:

a) Seal elevator shafts around cabs at floor stations.

Windbreaks

Minimal expense:

a) If open space is available, consider planting trees or large shrubs to act as windbreaks on the windward side of the building. The windbreak can have positive value on reducing wind impact, at least on lower floors. Trees and shrubs also can be used to reduce solar penetration, discussed elsewhere.

Heating and cooling

As already mentioned, heating and cooling together usually consume the largest single "block" of energy utilized by a building during the course of a year. In most cases, however, the heating/cooling system was designed only with initial costs in mind. As a result, the energy efficiency of the system seldom was a design criterion. Moreover, most systems were designed to meet extreme conditions which possibly could occur, but which seldom do. Accordingly, many are oversized and so perform in an inefficient manner.

There are many ways in which heating and cooling systems can be made more efficient. In general, these alternatives can be divided into five categories: operating practices modifications, maintenance modifications, systems modifications, control adjustment and modifications, and heat recovery.

Operating practices modifications (general)
Significant heating and cooling energy savings can be achieved simply by modifying the manner in which heating and cooling systems are operated. Several facts should be emphasized strongly, however.

- Operational savings are limited. While changes in operational procedures can save energy and energy costs, the net amount represents but a fraction of the potential which can be saved through other, more substantive measures. In other words, do not use operational savings as an excuse not to undertake other, potentially more beneficial measures.

- Operational savings will, in some cases, cause minor deviation from accepted standards of comfort. Some deviations may be more noticed than others.

- Professional assistance and guidance should be obtained before instituting any significant operational change. As an example, in certain situations setting a thermostat to 68°F during the heating season can cause the cooling system to be activated, wasting more energy than is conserved. (This situation was documented recently in a 25-storey midwestern office building at a cost of $ 1000 for three hours of winter cooling.) Likewise, setting thermostats at 78°F during the cooling season can bring on heating or reheat.

- Remember that each building comprises a unique situation. As such, the guidelines outlined herein should be recognized as general only. Each must be tailored for the building involved, preferably with professional guidance.

Here follow some of the guidelines for the more effective changes which can be made in operational routine to effect energy savings:

Minimal expense:

√a) Reduce use of heating and cooling systems in spaces which are used infrequently or only for short periods of time.

√b) Heat office building to 68°F when occupied, to 60° when unoccupied. This does not mean that air should be cooled if the temperature exceeds 68°F. Interior office spaces tend to experience significant heat gains due to lighting, equipment and people. Systems serving most areas use a combination of recirculated inside air and some outside air. As a result, the temperature may tend to stay at or above 68°F.

√c) Preheat building so that it achieves 65°F by the time occupants arrive. Complete warm-up during the first hour of occupancy. Lighting, people and use of equipment will aid in warm up.

√d) Turn heat off during last hour of occupancy.

√e) Cool office building to 78°F when occupied. Do not utilize mechanical cooling when unoccupied. Special consideration, however, must be given to computer rooms. Generally, the primary criterion is a constant temperature/humidity relationship. Manufacturers should be contacted to determine permissible ranges. Cooling usually is required when the equipment is operated, but should not be used to lower room temperature below the range of 78°F–80°F.

√f) Begin precooling operations so the building is 5° below outside air temperature, or 80°F, whichever is higher, by the time occupants arrive. Complete cool-down during the first hour of occupancy.

√g) Maintain retail store sales areas temperatures at 68°F during the heating season and at 78°F during the cooling season.

√h) Isolate storage room areas from sales area. Maintain storage areas at 60°F or lower in winter.

√i) Shut off all heating in garages, docks and platform areas.

√j) Consider closing outside air dampers during the first and last hours of occupancy and during peak loads.

√k) During cooling season evening and night hours, flush the building with cooler outdoor air.

√l) Allow natural humidity variations from 20% R.H. to 60% R.H. in occupied spaces.

√m) When appropriate, consider closing supply registers and radiators and reducing thermostat settings or turning off the electric heaters in lobbies, corridors and vestibules.

√n) Where sill height electric heaters are used, adjust thermostat so heat provided is just sufficient to prevent cold downdrafts from reaching the floor.

√o) Turn off humidifiers whenever the building is closed for extended periods of time, except when process or equipment requirements take precedence.

√p) Curtail humidification for areas such as hallways, equipment rooms, lobbies, laundry areas, and similar spaces.

√q) Turn off portable electric heaters and portable fans when not needed or during unoccupied periods.

√r) Turn on self-contained units, such as window and through-the-wall units, only when needed. Turn them off when the space is to be unoccupied for several hours.

√s) In mild weather, lower the cooling effect by running room cooling fans at lower speeds.

√t) Turn of all noncritical exhaust fans.

√u) Turn off reheat in all areas during summer, except where equipment requirements necessitate humidity control.

√v) When the sun is not shining during the heating season, close interior shading devices to reduce radiation from body to cold window surfaces.

√w) Develop an after-hours equipment operation checklist for use by custodial and other building personnel as well as occupants who may use various spaces after normal periods of occupancy.

√x) Schedule operating and maintenance work during the daytime, if possible.

√y) Wear heavier clothing during the heating season and lighter clothing during the cooling season.

√z) Reduce internal heat generation as much as possible during the cooling season. Typical sources of heat generation include lighting, people, machines, cooking equipment, etc.

Significant expense:
aa) Adjust and balance system to minimize overcooling and overheating which result from poor zoning, poor distribution, improper location of controls, or improper control.

Operating practices modification (kitchen and cafeteria areas)
A variety of steps can be taken to effect more efficient use of energy in kitchen, cafeteria and other food-handling areas:

Minimal expense:
√a) Turn off infra-red food warmers when no food is being warmed.

√b) Inspect refrigeration condensers routinely to ensure that they have sufficient air circulation and that dust is cleaned off coils.

√c) Inspect and repair walk-in or reach-in refrigerated area doors without automatic closers or tight gaskets.

√d) Train employees in conservation of hot water. Supervise their performance and provide additional instruction and supervision as necessary.

√e) Avoid using fresh hot or warm water for dish scraping.

√f) Keep refrigeration coils free of frost build-up.

√g) Clean and maintain refrigeration on water chillers and cold drink dispensers.

√h) Reduce temperature or turn off frying tables and coffee urns during off-peak periods.

√i) Preheat ovens only for baked goods. Discourage chefs from preheating any sooner than necessary.

√j) Run the dishwasher only when it is filled.

√k) Cook with lids in place on pots and kettles. It can cut heat requirements in half.

√l) Thaw frozen foods in refrigerated compartments.

√m) Fans that cool workers should be directed so they do not cool cooking equipment.

n) Consider using microwave ovens for thawing and fast-food preparation whenever they can serve to reduce power requirements.

Maintenance modifications

The importance of good maintenance to a program of energy management cannot be overemphasized. Not only will effective maintenance help ensure efficient operation of equipment and systems, but it also will help prolong the usable life of equipment.

The maintenance guidelines presented herein all should be performed to at least bring systems up to efficiency. They also should be continued on a regularly scheduled basis depending on the nature of your system, frequency of operation, etc. Inspection of many of the items mentioned also will provide you with some idea about the effectiveness of the maintenance program now in effect and the condition of your equipment, some of which may need adjustment, repair or replacement.

Realise that these guidelines are general only. Wherever possible, the manufacturer of the equipment involved should be contacted to obtain pertinent literature describing the maintenance procedures suggested. Otherwise, those who regularly install such equipment, or who design heating and cooling systems, should be asked to prepare manuals or guidelines.

Fans, pumps and motors. Proper maintenance of fans, pumps and motors can greatly improve operational efficiency and so eliminate unnecessary energy consumption. The following maintenance guidelines are suggested.

Minimal expense:

i) Fans

√• Check for excessive noise and vibration. Determine cause and correct as necessary.

√• Keep fan blades clean.

√• Inspect and lubricate bearings regularly.

√• Inspect drive belts. Adjust or replace as necessary to ensure proper operation. Proper tensioning of belts is critical.

√• Inspect inlet and discharge screens on fans. They should be kept free of dirt and debris at all times.

√• Inspect fans for normal operation.

APPENDIX C. SAMPLES OF ENERGY AUDIT SURVEY CHARTS

Building _____   Year _____

Month	Heating Deg. Days	Cooling Deg. Days	Electricity						Purchased Steam						Oil			Fuel Check Gas □ Coal □ Other □			Fuel/ Deg. Days	Total Energy Cost
			KWH	KWH/ Deg. Days	KW Demand		Cost		M (lbs.)	M (lbs.)/ Deg. Days	lbs/hr	Demand	Cost		Quant. (Gal.)	Cost		Quant.	Cost			
					Actual	Billed	Total	Per Ut.			Actual	Billed	Total	Per Unit		Total	Per Unit		Total	Per Unit		
1	2	3	4	5	6	7	8	9	10	11	12	13	14	15	16	17	18	19	20	21	22	23
Jan.																						
Feb.																						
March																						
1st Quarter																						
April																						
May																						
June																						
2nd Quarter																						
July																						
Aug.																						
Sept.																						
3rd Quarter																						
Oct.																						
Nov.																						
Dec.																						
4th Quarter																						
Total Per Year																						

Building Data

Gross Area (ft)² _____

Gen Notes: _____

_____

_____

Annual Energy Consumption In BTU's

Conversion Fac. 3413

	Quantity		BTU/Yr
1. Electricity	_____	kwh	_____
2. Purchased Steam	_____	(M) lbs	_____
3. Natural Gas	_____	MCF	_____
4. Oil	_____	Gallons	_____
5. Other Fuel	_____		
6. Total			_____

Energy Utilization Index

$$\text{EUI} = \frac{\text{Total Energy Consumption BTU's/yr}}{\text{Gross Area (ft)}^2}$$

= _____ BTU's/ft²/Yr

Seasonal Consumption (Hourly)	Connected Load	Maximum Demand	Winter	Spring/Autumn	Summer	Holiday/Weekend	Shutdown/Maint.	Annual Totals	Load Factor
Annual Schedule Hours									
Electricity – kW									
Lighting – internal									
Lighting – external									
Building Services									
Business machines									
Computer									
Production machines									
Other :									
Annual Totals									
Steam – lb(kg)/hr									
Space heating bar									
Production bar									
Amb. Sensitive process bar									
Annual Totals									
Hot Water – gal.(litre)/hr									
Space heating °F/°C									
Production °F/°C									
Amb. Sensitive process °F/°C									
Annual Totals									
Fuels and Water									
Gas cfh									
Oil 3500 secs /hr									
Oil secs /hr									
Oil 35 secs /hr									
Solid Fuel lb(kg)/hr									
Other									
Water gal(l)/hr									
Annual Totals									

Building or Site :

Zone or Plant :

Survey Date :

Survey Schematic

Checklist

Flow/return entry
and exit to zone.

Energy Sources :

Electricity
Steam
Hot water
Cold water
Compressed air
HVAC air
Fuels

Metering Facilities

Existing (M)
Space available (S)

Leaks Observed (L)

119

TYPICAL ANNUAL SCHEDULE OF DAY TYPES

Day Type	Winter	Autumn & Spring	Summer	TOTAL
Saturdays	12	15	19	46
Sundays/holidays	17	26	41	84
1st Production Days	14	17	19	50
Production Days	47	64	74	185
Total Days	90	122	153	365
Months	3	4	5	12

APPENDIX D. OUTLINE FOR AN ENERGY AUDIT CONSULTANCY

Brief
To set up and develop, co-ordinate and direct a consultancy service in total energy utilisation = energy audit and planning.

Objectives
Short term: to provide a service, complementary to existing mainstream consultancy services, to advise clients on optimum use of their available energy resources.

Medium/long term: to develop progressively by experience and by broadscale immersion in the subject so that the depth of expertise in energy utilisation is established beyond doubt at all levels of involvement.

Aims
1. Excellence in quality of service.
2. Total energy evaluation.
3. Satisfied clients.
4. Social usefulness.
5. Straight and honourable dealings.
6. Good reputation and influence.
7. Satisfied staff of reasonable prosperity.

Means
1. Qualified staff of high quality.
2. Efficient organisation.
3. A self-supporting operation in solvency terms.
4. Unity and enthusiasm for the total energy approach.
5. Leadership in the field of technical innovation in energy utilisation.

The concept of total energy utilisation
1. Evaluate load profiles and seek the balance — energy audit.
2. Gather together technologies and areas of involvement in order to balance loads and achieve maximum economies.
3. Attack waste in all its forms, and seek means of re-use.

The market scope
The market in energy utilisation studies appears to be extremely diverse, unless codified in some way to demonstrate a repeatability of problems. While this obviously requires a deeper study, some possible areas for development are immediately apparent.

Preliminary studies

Electricity usage
In the majority of cases, economies could be demonstrated in client's use of electricity by:

    a) better selection of tariff,
    b) improved selection and maintenance of light fittings and other electrical equipment,
    c) addition of power factor correction equipment.

The changes necessary are both simple and relatively low in capital cost.

Reduction in energy demand
Perceptive investigations into the construction and operation of the client's buildings, process and operation to ensure that energy savings are maximised. These would include:

    a) insulation of buildings, reduction in glazing losses, perimeter losses, eg, loading doors,
    b) process insulation,
    c) interchange of energy processes,
    d) waste heat recovery,
    e) reduction and re-use of wastes of all types,
    f) transportation and mechanical handling (energy used).

Choice and purchase of fuels
1. In a similar manner to electricity tariffs, there are a variety of fuel purchase deals available.

2. These 'tariffs' may include use of different fuels at various times of the year.

3. The fuel situation changes and alternative fuels can be evaluated on an economic basis, especially if all relevant factors (transport, storage, maintenance) are taken into account.

4. Under current disturbed industrial conditions, which seem likely to continue for some time to come, availability of at least one alternative fuel is an obvious insurance towards continuation of the client's operation, and expenditure of additional capital to this end could be justified.

<table>
<tr><td>Advanced studies</td><td>

Energy load balance
1. Balancing the energy requirements of a process, plant, building complex, or locality to make maximum use of recovered waste energy (usually heat).

2. Evaluation of the advantages of some form of energy storage against the load profiles.

3. Investigation into different operating methods (eg, two shift working) in order to correct load balance or to maximise efficiency of energy use. This could include fundamental changes in industrial procedures, for example, a measure of recycling waste metals instead of extraction from ore.

On-site generation
Evaluation of on-site generation with heat recovery versus standby plant under various conditions of client's operation and process, including a wide range of partial solutions.

Neighbourhood studies
Investigations on the surroundings of the client's plant/process to discover whether:

1. Waste energy is available at low cost from other sources.
2. Waste energy is saleable to other users.
3. Collaboration between public utility companies and the client could show mutual benefit, eg, generating plant with heat recovery, for use part by client and part by National Grid.

</td></tr>
</table>

Attack on the market

It is suggested that the service should be aimed at the following categories of client, in the given order:

Existing clients
At least the preliminary studies on reduction in energy demand and electricity usage should be offered to every client.

Industrial survey
Where there is (a) an obvious case of high energy use, or (b) a critical problem. Examples: (a) the glass industry; (b) food producing industries.

Governmental and Local Authorities: hospitals, developing universities, airports
Initially, it is recommended that the existing client situation is likely to provide a comparatively smooth entry into the market. By offering the service of preliminary studies, a relatively easy justification should be possible, thus generating credibility as a launching pad for wider endeavours.

In order to achieve the required results at the governmental level, it may be necessary to challenge established procedures and statutes — legislative lobbying may be needed, and may not be undertaken lightly.

Marketing the service
1. In-house contacts with existing and new clients.
2. Attendance at, and contribution to national and international conferences.
3. Preparation of articles in technical journals, and in specialist situations in national news media.
4. Personal approach to possible clients in selected industries.
5. Contact with governmental bodies, national associations in the selected industries.

Products and skills

– Ideas.
– Reports – engineering and economic.
– Drawings, specifications, schedules.
– Supervisory skills, including testing and commissioning.

Related technologies
1. Total energy, energy recovery and interchange.
2. Refrigeration, air conditioning, drying.
3. Oil and gas engines.
4. Electrical generation.
5. Incineration and waste systems.
6. Electrical distribution and use.
7. Natural energy sources: solar, water, wind.
8. Construction/civil/structural.
9. Mechanical handling, transportation.
10. Fluid flow and piped systems.

Personal skills
1. Open minded approach.
2. A full understanding of energy interchange, load balance, system studies.
3. Analytical techniques with practical intent.
4. Small-scale computer studies.
5. Cost effectiveness analysis on plant and process.

Recommended initial in-house staff skills
1. Electricity tariffs, distribution, use.
2. Total energy technology.
3. Fuel technologist/chemical engineer.
4. Engineering economist/accountant.

External liaisons
1. Academics in universities and research bodies.
2. Appropriate professional institutions.
3. Related product manufacturers.

Facilities required
1. Information retrieval (librarian) readily available.
2. Mechanical computation equipment, preferably intelligent terminal, such as Hewlett Packard 9830.
3. Expert typing and reproduction.

Principles of operation

Any commitment should be studied in terms of:

a) capability to do a professional job.
b) staff and techniques required.
c) investment in time, space and people.
d) involvement at the earliest possible stage of client contact.

On being better than others

1. Effective research = homework done well.
2. Dynamic (as distinct from passive) brief collection.
3. Professional attitudes.
4. Obsessive systems analysis = never be satisfied with a partial solution.
5. Multi-disciplinary approach.
6. Planned approach to each project = use of a defined plan of work.
7. Offer the service on a concept which the client can understand.
8. A wide view without losing sight of the main problem.

Profitability

Whilst a reasonable degree of profit ensures continuity of the operation, it may be necessary to invest for the future at early stages in the process of selling the concept. Concentration on entering the market by small contributions for existing clients is considered an essential counter.

Chapter 16

The operator's viewpoint

R Robinson

This chapter is based mainly on experience in a factory designed and built between 1968 and 1972, before energy conservation became a major issue. The points raised are those of problems with rather than successful design features of the unit, thus the chapter in isolation has a destructive slant towards the factory, which is not a true picture as the factory is a very successful production unit.

In order to give a proposed energy-economy measure any chance of success it must be totally acceptable to the users of the building, not just to the person paying the bills. It must, therefore, be presented to all affected by it in such a way that its success is desired by them. An imposed system has very little chance of succeeding, no matter how good it is. Thus advantages to all concerned must be highlighted, any restraints to be imposed must be recognised and made acceptable to the user and all objections by the user must be seen to be recognised as real and accounted for to his satisfaction. Production personnel are not generally interested in energy economies. They are very remote from costings and feel no personal involvement or concern for energy economy in the work place. Any changes made to their work procedure or environment are not seen as constructive in any way and are used as the scapegoat for any subsequent production problem, even when these changes have been thoroughly 'sold' to them, or better still elicited from them by consultation. In a conflict of opinion in this area, Production will usually win, as the cost of late or undelivered orders and subsequent loss of trade would normally far exceed the payback of any energy savings that could be made.

For example in one particular factory all 120 machines are connected to a central vacuum cleaner, and as each machine requires less than 30 minutes use of the system each shift, it is theoretically possible to run this plant for only 30 minutes each shift. However, production is maximised and better use is made of the operator's time if he is allowed to clean the machine when most convenient to him during the shift; thus the plant must be sized to cope with all machines calling together and run for the entire shift.

It is theoretically possible, and provision was made in the design of the building, to shut down about a third of the air-conditioning plants during non-production time, but, in order to prevent fresh air blowing back down the return air ducting, causing loss of conditions when in this state, the dampers in the return air-ducting of the shut down plants need to be closed. This action necessitates a visit to each plant by an operator. Another practical problem involved in this operation is that, when running, the plant's internal pressure is depressed below atmospheric, and any water or steam leak, or carry-over from the spray chambers, is contained within the plant casing until it is shut down. When the internal pressure returns to atmospheric and the water runs out, the operator's presence at this time helps to minimise any water damage to the production floor below. The presence of an operator is also desirable at start-up in order to check for a successful start and a satisfactory mode of operation, thus for a 9 hour overnight non-production period it is just about possible for one man to get through the complete procedure provided he does not take too long for a cup of tea between completing the stopping and commencing the restarting — always assuming that if this procedure is carried out the energy saving made will pay the man's wages.

The design brief for the factory called for large, freely interconnected, totally air-conditioned rooms. The size of the rooms necessitated multiple air-conditioning plants for each room, which brings its own problems of control of energy usage, in that some plants may be heating the room to compensate for heat losses to external walls and glazing whilst others in the same room are cooling to compensate for heat gain from machinery and people. Further, if the controls are not regularly and carefully maintained, especially as the air supply ducting runs for some plants are adjacent to the return air ductings of other plants, the situation can easily arise where the air-conditioning plants are busily conditioning each other and leaving the room to its own devices. A great deal of energy can be wasted in this sort of situation without anyone being aware of it.

A major problem which causes huge energy losses, but normally manifests itself as a people problem, is draughts. These arise from two sources:

1. Holes in the fabric, docks in particular. It was thought at the time of building that the slightly raised internal pressure of the building, created by the supply of air to the rooms being greater than the air returned from them, would negate any undue ingress of air through the docks etc. In fact, the wind does not need to be very far up the Beaufort scale, in an easterly quarter, before it is felt on the production floor. The building of a wall across the despatch docks and the provision of air locked access ways through this wall does not completely remove the problem, due to air pressure forcing doors open against self-closing devices. In other areas, staircases and lift shafts are the routes by which cold easterly air streams reach the production floors. This problem would be vastly reduced if it were practicable to raise and lower the roller shutter doors to each dock space each time a vehicle entered or left. However, it is an imperfect world and to the man on the docks, well wrapped up against the elements and involved in physical labour, it is not only unnecessary but far too tedious and distracting from his essential labours to bother about the doors, except to open them first thing in the morning and close them last thing at night. This problem is even more energy costly, but far less a people problem in summer. When the refrigerators are running to provide chilled water to remove moisture from the air and the people on the docks are enjoying being able to look out of the open roller shutters at sunshine, free exchange is allowed of conditioned air for unconditioned — overloading the plants and losing control of conditions in the production spaces.

2. The other cause of draughts within the building is that whilst on any one day it is possible to set a balance of air pressures between the various rooms such that there is no air transfer between them, this situation does not remain stable. Plants need to be stopped for cleaning and the quantities of air delivered by each plant are affected by the wind pressure on its intake and outlet grills, thus there is always air movement between rooms. This causes discomfort to those persons near to communicating doorways, which has to be combatted by the provision of unsightly screens or local heating to warm the air stream to a point where it is less uncomfortable. This movement of air also means the air-conditioning plants' return air sensors do not always 'see' the air from the same place, causing control difficulties.

During the design of the building the engineers involved wished to eliminate or vastly reduce the amount of glazing in order to remove the inherent problems it creates in air conditioning. They were, however, overriden by others and the glazing present, minimal though it may be, despite being double glazed and covered with a heat reflective film, causes problems in several ways. The heat gain and loss through it obviously result in higher fuel bills; additionally, fairly isolated hot dry areas can be formed by direct sunlight causing drying out of product and materials which is very bad for product quality. During cold weather condensation on the glazing is a problem despite energy consuming fans blowing air over it. People working near the glazing are subject to an unpleasant downpour of cold air; the air conditions adjacent to the glazing are obviously off spec if the rest of the room is correct.

In an effort to save fuel during the fuel crisis the lighting fittings over gangways and offices were disconnected, but once the feeling of virtue over saving fuel had worn off the factory looked so depressing with dark walls and holes in the lighting pattern, that all fittings were soon returned to service by popular demand.

Within the total energy plant it is more economical to run the station with sufficient turbine/alternator sets on the bars, such that the failure of one machine does not cause a cascade shutdown of the entire plant, as the half hour stoppage of production by one such outage per annum is far more expensive than the extra fuel burnt during a year by running the generating plant in this less efficient manner.

In the office block buildings equipment has been installed which will provide optimum start-up facilities and whilst is is confidently expected that up to 25% of fuel bills will be saved in these areas generally, there has not been the sort of weather this winter where much saving can be expected.

Chapter 17

The potential – how can we contribute best?

J Hampson

The opportunities for engineering skills are widened by the increased international awareness of energy conservation. There is even tremendous scope for the optimisation of the existing methods of generation, conservation and transportation of energy. In the long term future we can put our faith in windmills, hydropower, solar power, wave energy and tidal power. We can look towards converting the vast heat energy reserves in the outermost 2-plus kilometres of the earth's crust which exceed the equivalent heat energy of the earth's known reserves of fossil fuels. Geothermal stations are already planned for Japan, Mexico and Hawaii, which have numerous volcanoes – probably the most dramatic expression of nature's power.

This book is typical of most on the topic of energy. It is structured around building design but more space has been devoted to the specialist elements than to moulding our ideas and skills together towards achieving balanced designs, which are perhaps the key to our future. Energy cannot be properly debated in isolation without full participation. Architects were to a large extent ignorant of the serious future energy penalties which their designs imposed. The buildings were often made habitable by expensive and sophisticated engineering solutions. We must help architects to develop a new design approach which is compatible with the many complex requirements which include energy conservation. The designs will change with climate conditions, national economic policies and user needs. There is a need to appraise the extent of our involvement past, present and future.

Up to 50 years ago building structures were designed from load-bearing materials and the restrictions imposed on spanning automatically controlled window sizes. These buildings offered not only reasonable thermal insulation but also excellent thermal capacity. The development of structural engineering designs made frame construction a reality, leading to the ridiculous extremes in the post-war years of lightweight infill panels. The reduction in thermal capacity resulted in high energy demands for the majority of users. Taller buildings were built, which suffered not only from excessive glass areas but also from attendant problems of air leakage. Deep planned buildings were designed for restricted sites and improved capital costs per capita but necessitated air conditioning and hence even higher energy penalties.

We need to develop an awareness of the effect of the structural/building envelope concept on the creation of an artificial climate. There are many opportunities for considering alternative designs for structures and systems to suit a particular architectural concept or preferably different concepts. These opportunities include:

Orientation	:	Consider energy as well as planning and geotechnical aspects.
Thermal capacity	:	Lightweight capacity for primary schools and heavyweight capacity for most other buildings. This relates to load-bearing vs frame and cladding, GRP cladding or similar vs concrete cladding or similar.

Good thermal insulation without condensation	
Optimisation of space utilisation, eg, fenestration	: Compatible services/structural systems.
Windproofing as well as weathering	
Services/structural zone depths	: Coffered slabs for flexibility and beam/slab solutions for optimum overall structure services depth.
Plant configurations and influence on system choice and layouts	
Elevational treatment	: Thermal capacity, perimeter services, vertical distribution of services and stability.

By developing our engineering skills in this area we will be able to participate jointly with those enlightened architects who are now considering energy as an important design parameter. Similarly we will communicate better with external consultants and building owners, and so achieve improved design team collaboration and more harmonious design.

There are other areas of energy utilisation which are perhaps worthy of study, such as the comparable energy used in the production and erection of steel and concrete structures. However, energy is already in the true building costs and it would normally be wrong to complicate the decision-making process by elementalising costs in this way. However, where energy costs are escalating and they represent a large proportion of the total costs then such trends can affect design concepts. The inflationary effects on labour and materials have so far overshadowed this aspect.

Where we are able to participate more fully in building design, we can seize more opportunities in this area of our work. There will probably be fewer new buildings constructed annually, and an increasing number of restoration and refurbishing projects and these services make such work more meaningful in financial content. These opportunities have extended into several research and development opportunities embracing total engineering factors as well as pure system studies related to heat recovery.

Regardless of the present-day viability of any alternate energy source the government has responsibility to ensure that there is a full range of energy alternatives available for future use. Central government is already committed towards the researching of wave, wind and tidal power together with geotechnical and solar energy. Research projects will continue to be commissioned on an expanding basis. In a small way we have already been involved in this exercise. Many of us would like an opportunity to increase our participation in this sphere but we should examine what we really have to offer.

Generally speaking, those of us who are building designers in structure, heating/air conditioning, electrics, etc, have limitations in skills and experience and lack sufficient time to carry on with our present activities as well as enter this new field. Heavy mechanical engineering skills and experience are needed and the difference in education/training and attitude between a turbine designer and an HVAC engineer is greater than that between a structural engineer and a civil engineer. Therefore in order to offer a complete engineering service we would need a major intake of new skills. There will be ample opportunity to fully participate in these new areas of research with or without further diversification. With regard to diversification, the initial development exercise of the power industries such as the petrochemical industry has traditionally been manufacturer-design orientated. There is no evidence to suggest that this situation would change when new energy sources are exploited. Any intensified diversification of our skills in this area would require to be justified not only by new work, but also by a longer-term need for the new skills perhaps in associated fields. Furthermore, a lot of development work will inevitably be offered to the research establishments and universities.

Nevertheless our practice is unique in the range of skills which we have to offer. Since the research programme will be an expanding one there is a need to continue to use our imagination to examine ways of increasing our possible participation in such exercises. There seems little point in pursuing prospects which will not at least lead to the employment of our established services. Similarly the Ove Arup expression that good ideas must be tempered by realism means that there is little point in pretending that we have all the necessary skills, and hence there is a need to collaborate with others.

We have recognised our continued involvement in energy research work. Energy is only one of the many parameters for building design. As such it must be thoroughly researched but when it has been it is hoped that the overall design decisions will continue to be balanced ones. The majority of the buildings which we design are for people who must be made comfortable and the quality of their aesthetic/visual environment must be enhanced.

Summarising therefore:

1. We need to develop a greater awareness for the potential energy wastage created by our building designs. The design decisions have to be balanced against the conflicting requirements of heat, light and sound as well as visual and aesthetic parameters.

2. There are prospects for continuing to participate in research and development projects related to energy conservation and heat recovery in buildings only if we advance our in-house research and development work to complement our project skills and experience.

3. There are prospects for participating in the vast research and development programme for alternative energy sources. This could embrace our Structural, Building Engineering, Civil Engineering, and Central Services skills in part or full as required.

4. There are long-term prospects for participating in energy stations traditional and futuristic, but the comments of joint venture particularly apply in view of the large manufacturer/designer participation which has been traditionally recognised.

5. There will be continuing prospects for an extension of our Building Engineering Service into energy consultancy, which will increase the potential for all our services and provide better continuity in client contact.

Index

132

DATE DUE

UCR NOV 1 5 1981		
UCR NOV 1 5 1981		
OCT 24 1986		
FEB 29 1988		
MAY 20 1993		
MAY 17 2001		
GAYLORD		PRINTED IN U.S.A